Basic
structural detailing

Basic
structural detailing

R. Benton

Basic
structural detailing

Longman
Scientific &
Technical

Longman Scientific & Technical,
Longman Group UK Limited,
Longman House, Burnt Mill, Harlow,
Essex CM20 2JE, England
and Associated Companies throughout the world.

First published 1989
Second impression 1991

British Library Cataloguing in Publication Data
Benton, R.
  Basic structural detailing.
  1. Buildings. Construction. Technical
  drawings. Draftsmanship
  I. Title
  692′.1′0221

ISBN 0-582-41331-1

Set in 10/11 AM Comp/Edit Times

Printed in Malaysia
by Percetakan Anda Sdn. Bhd.,
Sri Petaling, Kuala Lumpur

Contents

Acknowledgements

We are grateful to the following for permission to reproduce copyright material:

Bison Concrete Ltd. (Southern Division) for fig. 9.4 from load/span tables; British Constructional Steelwork Association Ltd. for figs. 27.5, 27.7, 27.8 from *Metric Practice in Structural Steelwork*; British Standards Institute for data, figures and tables included in figs. 8.1, 8.2, 8.3, 11.3, 13.1, 13.2, 13.3, 14.2, 14.3, 15.1, 15.2, 15.6, 15.9, 16.1, 16.6, 16.7, 17.4, 21.4, 23.2, 23.4, 23.5, 23.6, 24.2, 24.3, 25.7, 27.2 (complete copies can be obtained from BSI at Linford Wood, Milton Keynes, MK14 6LE); CCL Systems for fig. 17.8c from their catalogue; Cooper & Turner for fig. 24.4c/d/e from their catalogue; Servicised Ltd. for fig. 9.4 from their catalogue.

Author's Acknowledgements

The author would like to thank the following individuals and companies for their help and advice during the writing of this book:

Adamson-Butterley; A.R.C. Building Ltd. (formerly Marples-Rigway Building Ltd.); Avon County Council, County Engineer and Surveyor of; British Board of Agrément; BARFAB Reinforcements; British Cement Association; British Steel Corporation; Building Research Establishment; Clarke Bond Partnership; Conder Group Services; Creteco Ltd.; Director, Dept. of Transport (SWRO): Fairfield-Mabey Ltd.; Goodhead Print Group plc (formerly Techomes Ltd.); Gwent County Council, County Engineer and Surveyor of; International Paints; Pearce Building Contractors; Norwest Holst Construction Southern Ltd.; Rom Ltd.; Rowecord Ltd.; South Western Regional Health Authority; Square Grip (Western) Ltd.; Steel Construction Institute; Thomas Steelwork Ltd.; York Trailers; Zinc Development Association.

Introduction

Structural drawings are the principal communication between designers and builders. They must be technically accurate, clear and unambiguous, not only to ensure structural strength of the completed works, but also because they are used as a basis for cost estimates and payments to the contractor.

The drawings bring together all aspects of the design process and in addition to developing drawing skills, technicians and engineers can broaden their appreciation of construction methods.

This book covers the requirements of the BTEC Structural Detailing II syllabus and also explains the background to detailing so that students can relate their studies to the industry in which they work. It will also be appropriate as a basic text for students who become involved in computer-aided detailing and as a general reference book in design offices.

While writing the manuscript, I received a lot of help from companies and organisations in the construction industry. A number of specialists read and corrected sections of the book for me and I was allowed onto numerous sites to take photographs. I have included what I hope is a full and adequate list of acknowledgements.

I would also like to thank all those friends and colleagues who patiently listened to my enquiries and offered constructive comments in my attempt to produce a consensus view of the art of structural detailing. In particular, Gordon Duffin who in addition to giving much considered advice, also prepared the reinforcement detail drawings in section 20.

Part 1

Section 1

Construction drawing

1.1 Introduction

The structures of building and civil engineering projects are built to designs prepared by structural engineers and technicians. Every project is different, requiring individual designs and construction drawings.

Producing the drawings is normally the principal responsibility of the technician, and although they must be prepared in accordance with recognised standards and codes, the technician should be able to develop his or her own individual drawing style and acquire an understanding of the principles of design.

Construction is broadly separated into building and civil engineering works.

Building work includes: hospitals, schools, offices, warehousing and factories.

Civil engineering includes: highway and railway bridges, power stations and similar heavy industrial works.

1.2 Building works

Building and civil engineering works are carried out using different forms of contract. Building construction uses the RIBA (Royal Institute of British Architects) conditions of contract with a JCT (Joint Contracts Tribunal) form of contract.

Several different procedures are used for the design and construction

4

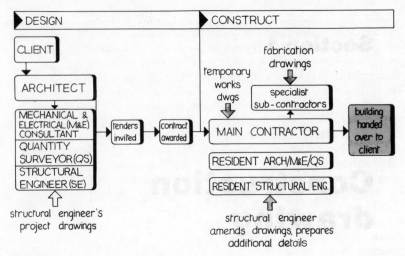

Fig. 1.1

of buildings. Figure 1.1 shows a simplified network for the traditional method including the stages when structural drawings are required.

A project normally originates from a brief which is basically a specification for the completed building. A preliminary brief is usually prepared by the client, who may be a property developer, housing association, hospital board, etc. It may later be modified in consultation with an architect. The brief includes items such as: the type of accommodation and the capacity for use; the floor areas in an office block; the number of beds in a hospital; etc.

Building design is normally undertaken by independent consultants who accept commissions. The client employs an architect to advise on requirements and to prepare designs for the building. On all but the smallest projects the architect uses the services of other specialist consultants. A structural engineer (SE) is required to design the structural framework of the building. A mechanical and electrical (M&E) consultant designs the heating, electrical and other services, and a quantity surveyor (QS) coordinates financial aspects of the work, preparing cost estimates and tender documents. These consultants are often appointed on the recommendation of the architect but are employed directly by the client. On building projects, the structural engineers are normally a firm of consultants led by chartered engineers.

1.3 Drawings for building works

Although the whole project team (architect, structural engineer, mechanical and electrical consultant and quantity surveyor) may work on the project simultaneously, the architect normally starts the design process with preliminary scheme drawings which show in basic outline, the plans, elevations and layouts of the proposed building. Working from these drawings, the structural engineer investigates suitable frameworks, preparing basic calculations and scheme drawings to show the layout of the structure, choice of materials (concrete, steel, etc.), floor spans, sizes of the principal elements and so on. The schemes are evaluated for suitability and cost, and a structural arrangement chosen. The architect and engineer are then able to proceed with their own detailed designs. The mechanical and electrical consultant supplies information about building services. The structural engineer's main interest in the services is usually the holes required in floor slabs to accommodate ductwork and items of heavy plant requiring support.

The preparation of working drawings and schedules follows. About 20 structural drawings may be required for a typical medium-sized office block. Building design normally follows a well defined routine and progress from preliminary schemes to full project proposals and detailed design usually runs smoothly. Although at the detailed design and working drawing stage the project team can proceed with their own designs independently, as drawings are produced they are circulated via the architect to the other consultants for information and comment. This continuous coordination is essential to the smooth running of a design programme. During this stage drawings are normally stamped 'preliminary'.

The pace of most building design and the need for continuous coordination between the members of the project team means that it is not always possible to have a complete set of working drawings and schedules ready at the tender stage. Instead, the engineer may use the principal working drawings together with estimates for items such as the quantities of steel reinforcement. Alternatively, the preliminary scheme drawings may be developed into suitable working drawings.

1.4 Design team

In the text the name 'design team' is used to describe the group of engineers and technicians who design a building or civil engineering structure. The precise responsibilities of each member of the team vary according to the skills of the individuals. In general, the design is controlled by a project engineer who carries out the preliminary scheme design and establishes a basic form for the structure. The design engineer then works with the other engineers in the team, preparing design calculations for the chosen scheme. The technicians prepare drawings and schedules and may also prepare calculations for some of the more straightforward elements in the structure such as lintels and stair flights. This would be done under the supervision of an engineer.

The size of the design team depends on the size and type of the project. A typical medium-sized office development may require a project engineer, assistant engineer and two technicians. Large projects such as hospitals, shopping complexes and arts centres, are often divided up and shared between small groups of engineers and technicians who are responsible to a coordinating project engineer. On a small job, a senior technician may work independently, doing both the design and drawing him/herself.

Early in a project the design team normally commission a site investigation to determine ground and foundation conditions. The work is carried out by specialist contractors who sink bore-holes, take soil samples, carry out tests and prepare a technical report. For a small project with lightly loaded foundations, it may only be necessary to dig trial-pits (usually up to 4 m deep) using an excavator, and then carry out simple tests and record the findings.

1.5 Specialists' drawings

Further structural drawings are required after the contract has been awarded and work on site has begun.

The main contractor may prepare drawings for the temporary site works including shoring and formwork. Details which relate to the stability of the works may be inspected by the structural engineer.

Structural steel frames are used extensively in building works. The steelwork is supplied by specialist sub-contractors (steel-fabricators) who have their own design and drawing offices. Working from the structural design team's drawings they prepare their own fabrication details which are sent to the engineer and architect for approval before the steelwork is made.

On site, the structural works are monitored by the consultant's resident engineer who approves the work in progress. During the contract, minor amendments are often required on the drawings and occasionally additional details are required. The design team undertake this work. For building works, the architect is the client's representative and is responsible for issuing variations and additions to the contract. This is done with an architect's instruction (AI). Although the structural design team may issue new information directly to a contractor, it must always be confirmed with an appropriate AI.

1.6 Civil engineering works

Civil engineering works use the Institution of Civil Engineers (ICE) form of contract. A simplified design and construction network for an ICE (civils) contract is shown in Fig. 1.2. Like the JCT form, the main contract is made between the employer (client) and the contractor, but, in this case, the client commissions a consulting engineer to design and

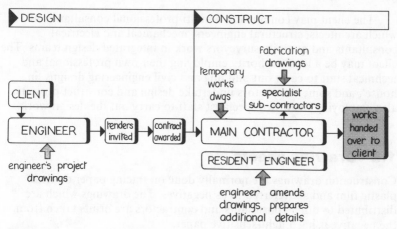

Fig. 1.2

supervise the works. The engineer is normally the only consultant, although an architect may advise on visual aspects of a large and environmentally sensitive project.

Civils projects often involve considerable liaison with authorities such as gas, water and drainage services; manufacturers of mechanical plant and so on. Otherwise the project design proceeds from preliminary schemes to working details in much the same way as a building contract, with design teams of engineers and technicians preparing calculations and drawings. Full working drawings are normally prepared as part of the tender documentation.

The main contractor may prepare temporary works drawings, and if the project includes structural steelwork, a steel-fabricator is employed who also produces fabrication drawings.

The work on site is monitored by the consultant's resident engineer who checks the work in progress and authorises payments. On a very large project, a small design team may be established on site to make amendments to the working drawings and prepare special site drawings. The work may be concluded with the preparation of 'as-built' drawings which are the original working drawings updated with all the amendments and additions required as the work progresses. As-built drawings are needed for the continuous maintenance of the completed structure.

1.7 Alternative procedures

The networks show project teams composed of separate consultants commissioned to work together on a contract, but building and civil engineering design may be undertaken in other ways.

The client may commission a multi-professional consultancy in which architects, structural engineers, mechanical and electrical consultants and quantity surveyors work in integrated design teams. The client may be a local authority employing their own professional and technical staff to carry out building and civil engineering designs 'in house', and some contractors undertake design and construct projects, also employing their own specialist staff to carry out the design work.

1.8 Drawing sheets

Construction drawings are normally done on tracing paper or clear plastic film and are referred to as negatives. The drawings which are distributed to other consultants and contractors are prints taken from the negative using a light-sensitive paper.

For the engineer, much of the preliminary design is often intuitive and he/she may choose to do his/her own scheme drawings to get a 'feel' for the structure. Alternatively, a technician may work with the engineer to produce this preliminary information. At this stage, the presentation of the proposed structural framework as a simple scheme drawing is more important than achieving an excellent drawing style, and the scheme drawings are often done in pencil on plain tracing paper, allowing changes to be made quickly and easily as the design evolves. The philosophy 'if it looks right, it is right' is valuable guidance at this stage.

As the design advances, the working drawings are done on standard-sized cut sheets which are pre-printed with a border and panels for titles, drawing numbers and similar information. The sheets may be good-quality tracing paper or clear plastic film. Tracing paper is economic but is liable to crease and tear if it is not handled carefully. Plastic film is more expensive but gives a smooth hardwearing surface and does not tear. Working drawings are usually done in ink, occasionally in pencil.

1.9 Typical drawing sheet

A typical pre-printed drawing sheet used by the structural design team is shown in Fig. 1.3. The border prevents the linework of the drawing getting too close to the edge of the sheet, important since the negatives may get damaged at the edges and prints used on site deteriorate rapidly, fraying at the fold lines and around the edges.

The border may also be marked up with scales for microfilming.

The panel on the right-hand side of the sheet includes provision for:

Name of the engineer: normally pre-printed on the sheet.
Name of the client or architect: written onto individual sheets.

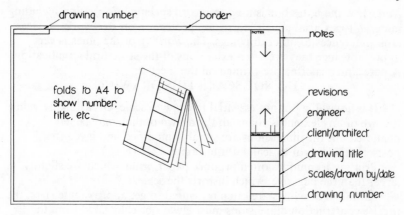

Fig. 1.3

Title: Identifies the project and particular drawing clearly, concisely, and in large letters (typically 7 mm high).
For example: BURNT MILL SPORTS CENTRE
                    FOUNDATION LAYOUT
Scales/drawn by/issue date: indicates the scales used on the drawing (1:100, 1:50, 1:20, etc.), the name or initials of the draughtsperson and checker, and the date of issue.
Drawing number: drawings are individually numbered for identification. The number normally consists of the project identification/individual drawing number/revision.
Project identification: The project is usually identified by a number or combination of numbers and letters. Consulting engineers working on building projects often use a four-digit system, a highway project may incorporate an appropriate road classification or identify the structure with a number derived from the Ordnance Survey coordinates of the site. Similar zoning identifications are used for other types of structure. Ideally the project identification number should be concise.
Individual drawing numbers: may run in sequence 1, 2, 3, etc., or use blocks of numbers for different sections of the work, 100, 101, ..., 200, 201, ...
Revision: when a drawing is amended it is given a sequential revision letter A, B, C, ... and the drawing is re-issued.
    Typical drawing numbers would be:
                    5521/12 rev C.
                    78128/23 rev A.
Revisions box: this is used to summarise details of the amendments and includes 'revision', 'revised by' and 'date' columns. In addition, circles may be drawn on the back of the negative to highlight the revised details, but this is not common practice.

10

Notes box: the notes box is used to record specific information including material types, trial panels, corrosion protection on steelwork, services, cross-references to other drawings. The wording of the notes is very important since they are often extensions of the structural specification. A note which is often pre-printed on the sheet is:

DO NOT SCALE THIS DRAWING

It is advisable but not essential to draw to scale. Occasionally, when the size of an element is changed, the drawn outline may be left unaltered but the dimensions are suitably modified and given the postscript NTS (not to scale), for example: 1710 (NTS).

Printing techniques often produce paper prints which are slightly larger than the negative, which distorts the scale.

Although drawings are often measured to get approximate sizes, the precise construction dimensions must always be calculated from the actual dimensions given.

1.10 Alternative drawing sheet

Specialist fabricators may use a drawing sheet similar to the type shown in the preceding section. Alternatively they may use the typical layout shown in Fig. 1.4. The sheet uses a border and identifies the name of the fabricator, the project title, notes, drawing number and revisions. The notes are normally less extensive and many are standard details about welds, surface treatment and so on.

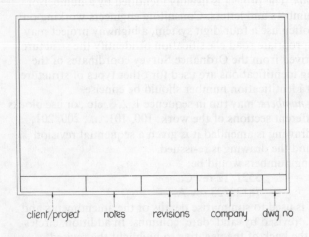

client/project   notes   revisions   company   dwg no

Fig. 1.4

1.11 Drawing sheet sizes

The pre-printed sheets used for construction drawings are normally of a standard size chosen from the ISO (International Standards Organisation) A or B ranges for paper sizes. Sizes between the standard A1 (594×841 mm) and B1 (707×1000 mm) are most common. Smaller sheets, for example A2 (420×594 mm) may be used for details, but design teams normally prefer to use one sheet size for all drawings on a project. Sheets much larger than B1 can be difficult to handle—particularly on a windy site! During the design and detailing phase, the negatives are usually stored flat in a plan-chest. On completion they are hung in metal cabinets using a purpose-made card or plastic hanging strip.

Design teams and contractors may also use A4-sized (210×297 mm) 'sketch' sheets which are blank tracing sheets pre-printed with a simple heading. These sheets are very useful for issuing additional information as the contract proceeds, and they are often given an 'sk' number, for example 4031/sk4.

1.12 Printing

The drawings distributed among consultants and contractors are paper prints made from the negatives. These are normally produced by the dyeline process which uses plain paper coated with a light-sensitive material. The prints usually have black lines on a white background. The popular image of engineering drawings as 'blueprints' (white lines on a blue background) has not been appropriate for many years. A4-sized sketch sheets may be reproduced as dyeline prints or photocopied.

The dyeline process can also be used to make copy negatives by printing onto a transparent plastic sheet coated with a light-sensitive material. It is also possible to reproduce prints of an existing paper print. 'Prints of prints' are usually taken as record copies of drawings for which no negative is readily available—sub-contractors drawings, etc. The copy is usually very dark and patchy, but may be adequate as a record of important information.

Section 2

Drawing instruments

2.1 Introduction

This data sheet describes basic drawing instruments used to prepare construction drawings.

A drawing may be worth several hundred pounds in labour costs, and high-quality instruments and materials are essential to prepare drawings to the standards required in the construction industry. An extensive range of suitable equipment is available through drawing-office suppliers, and manufacturers are continually developing and extending their product ranges.

An engineer or technician who draws regularly normally has a personal set of basic drawing instruments including a set square, scales, pens, pencils, erasers and simple templates. In addition he or she may have access to office sets of more expensive and less-used equipment, such as special curve templates, compasses and electric erasers.

2.2 Drawing boards

The parallel-motion board is commonly used in construction drawing offices (Fig. 2.1(a)). It is usually made from dense chipboard with a veneered or plastic finish. The straight edge is plastic or hardwood and metal with plastic edges and is attached to a simple mechanism of linked pulleys, wire loops and balance weights which keep it parallel as it is moved across the board. The board may be mounted on metal legs (b)

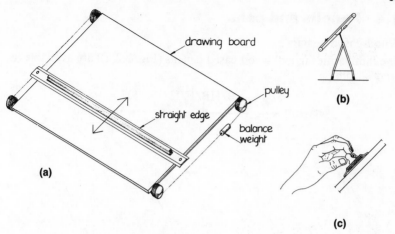

Fig. 2.1

or a frame bolted to a table. The angle and position can be adjusted through a quick-release mechanism and the balance weights on the parallel motion prevent the straight edge from slipping, even when the board is set at a very steep angle. The grip is bevelled and can be used as a shelf for pens, pencils and other drawing instruments (c).

2.3 Line thickness

On engineering drawings, different line thicknesses are used for different parts of the work.

Generally, the principal outlines and large sections are drawn with thick lines, details and smaller-scale outlines use medium lines, and dimensions and hidden details use fine lines. The use of a range of line thicknesses gives immediate visual emphasis, depth and interest to the drawing.

Technical pens and certain types of pencil are designed to draw lines of specific thickness in accordance with international standard recommendations. The ISO line thicknesses, which are also colour coded, are:

0.13	0.18	0.25	0.35	0.50	0.70	1.0	1.4	2.0
violet	red	white	yellow	brown	blue	orange	green	light grey

The line thicknesses increase in the ratio $1:\sqrt{2}$ ($0.13\sqrt{2} = 0.18$, $0.18\sqrt{2} = 0.25$, $0.25\sqrt{2} = 0.35$, etc.) which is the same as the ISO drawing sheet sizes.

2.4 Pencils and pens

Wood-cased pencils

The familiar hexagonal wood-cased pencils (Fig. 2.2(a)) are available in grades from:

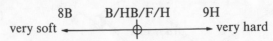

8B   B/HB/F/H   9H
very soft ←――――――⊕――――――→ very hard

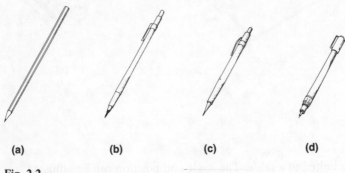

(a)   **(b)**   **(c)**   **(d)**

Fig. 2.2

Coloured pencils are also used on construction drawings for shading. In the dyeline process, different colours print to different tones. Green and blue print to a mid-grey, and yellow prints almost black.

Clutch pencils

The body of a clutch pencil (b) is normally plastic and metal. The leads are replaceable and are held in a spring-loaded collet. The lead is released by depressing the button at the top of the pencil, allowing it to be advanced as required, or retracted to protect the point. The button is detachable to allow replacement leads to be inserted. The grade of lead used can be easily identified using interchangeable colour-coded plastic buttons which are supplied with boxes of leads.

Automatic pencils

Automatic pencils (c) use replaceable leads and are designed to draw lines of specific thickness. Pencils are made for line thicknesses of 0.3, 0.5, 0.7 and 0.9 mm, and are colour coded for easy identification.

Technical drawing pens

Most ink drawing is done using technical drawing pens (d), which produce a line of fixed width. Nibs are made to the standard range of ISO line thicknesses and parts of the pen such as the sleeve and cap are appropriately colour coded. Special inks are used in technical pens and compass attachments and other accessories are available.

Technical pens are also used in computer drawing equipment.

2.5 Erasers

Special erasers are used with the tracing sheet and film used for construction drawings.

Pencil lines are easily removed from tracing sheet using a soft vinyl eraser (Fig. 2.3(a)).

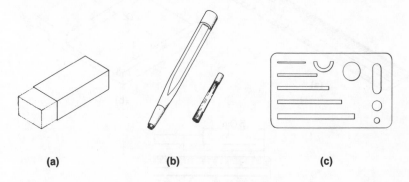

(a)   (b)   (c)

Fig. 2.3

Ink lines can be removed using solvent-loaded vinyl erasers. Fibreglass erasers (b) use replaceable inserts of fibreglass hairs which are very abrasive. Ink can also be removed by scraping with a sharp razor blade. This last technique is particularly useful for removing very small areas; however, it may damage the paper. The more severe methods of erasing destroy the paper's surface. Ink lines subsequently drawn on the area spread and may be indistinct. The problem can be reduced by polishing (with a smooth pen-cap for example) which partially restores an effective drawing surface.

Drawing film is transparent plastic sheet coated with a lacquer. Pencil and ink lines should only be removed using the correct type of eraser since, if the lacquered surface is broken, it is not possible to draw on the base plastic sheet.

The erasing shield (c) is a thin metal plate perforated with holes of different shapes and sizes. It is used to mask the drawing adjacent to a specific area which is to be erased.

For clearing large areas, special automatic (electric) erasers are available which use replaceable rubber eraser inserts. Special cleaning fluids are also used.

2.6 Scales

For construction drawings a scale rule is used to measure full-sized
dimensions directly. The typical double-sided civil engineering rule
(Fig. 2.4(a)) is divided on the four faces, each face carrying two scales.
The scales range from 1:5 to 1:1000. In the example (b), the distance
from the slab edge to the step is 5 metres. For a 1:100 drawing, this
dimension can be measured directly using the appropriate dividings (c).

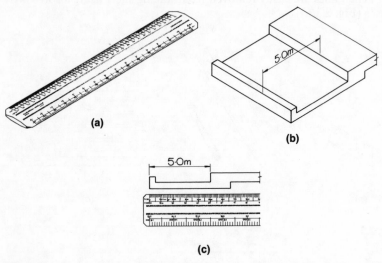

(a)

(b)

(c)

Fig. 2.4

· Other scales are available with dividings up to 1:1250 and 1:2500 for
use on land survey drawings.

2.7 Set squares

A set square is used against the tee-square or straight edge on the
drawing board, to draw vertical or angled lines (Fig. 2.5). Set squares
are normally made from clear or tinted plastic material and may be
(a) 45°, (b) 60° or (c) adjustable. Of the two rigid squares, the 45° is
probably more useful since this angle often occurs in construction
details, for chamfers and so on, but drawings are normally done with the
adjustable square which is basically a 45° square which opens on a
lockable protractor scale (inset). Squares are made with (d) bevelled or
(e) plain edges.

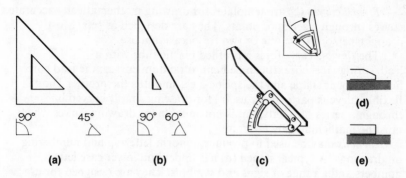

Fig. 2.5

2.8 Templates and stencils

A wide range of templates and profiles are used for construction drawings (Fig. 2.6). The *circle template* (a) is a rectangular sheet punched with small diameter holes in increments from (typically) 1 to 35 mm. Centre-line cross-hairs are marked on each hole for accurate location. Similar templates exist for squares, hexagons, elipses and technical symbols.

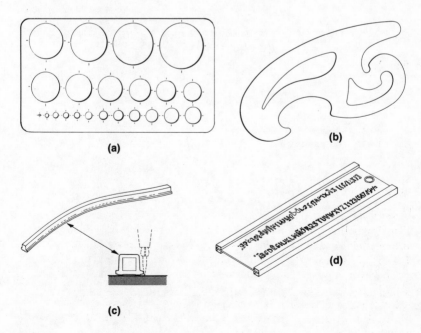

Fig. 2.6

French curves (b) are templates for drawing mathematically accurate curves through a series of points. They are designed as sets. Most draughtspersons have one or two for occasional use.

The *flexible curve* (c) is a profiled plastic tube with a lead/spring-steel core. It can be bent to random curves; the lead core holds the bent shape and the spring-steel smooths the profile. The flexible curve is particularly useful for drawing simple curved lines through a series of points, but is not suitable for drawing curves which must be mathematically accurate.

Stencils may be used to produce uniform lettering and numbering on drawings. A typical stencil (d) has upper and lower case letters, numbers and a range of signs and symbols. They are designed for use with technical pens.

Stencilling should not be regarded as a substitute for the practised freehand printing which gives a drawing individuality. However, even skilled draughtspersons often use stencils for the large letters required in headings and title blocks since these are particularly difficult to form neatly.

Section 3

Projection and setting out

3.1 Introduction

The design team's drawings consist of views of a structure set out to show its exact shape and size. The three-dimensional structure is shown as a series of two-dimensional aspects (projections), which are chosen and arranged according to standard conventions. This section describes the projection methods most commonly used on construction drawings.

3.2 Definitions

The following are the main terms used in the descriptions of projections:

Plan—a view of an object taken from above.
Elevation—an external view of the vertical face of an object.
Section:—a view taken by making an imaginary cut through an object and drawing the cut face.

Figure 3.1 shows a rectangular block with an offset hole. It is viewed from three positions:

The view from position (a) shows the plan.
The view from position (b) shows an elevation.
The view from position (c) shows a section.
The dotted lines on the elevation (b) show the hidden detail of the internal hole.

20

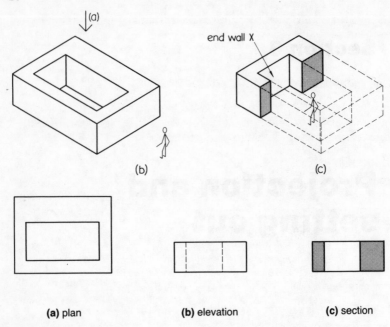

(a) plan   (b) elevation   (c) section

Fig. 3.1

Cross-sections may be drawn to show only the cut face or the cut face and the details immediately beyond. The section (c) is positioned to include the elevation of the end wall X. Sections should be taken perpendicular to the axes of the structure and the positions chosen to include as much information as possible.

The views are drawn to scale.

3.3 Projection methods

The principal methods of projection are shown in Fig. 3.2. Orthographic

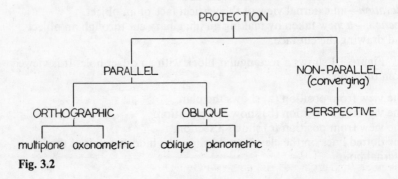

Fig. 3.2

and perspective are the commonly used methods of projection on construction drawings.

3.4 Multiplane projection

For this method, the structure is viewed 'square on' to give a series of plans, elevations and sections. There are two basic types: first-angle projection, and third-angle projection.

First-angle projection

Figure 3.3 shows a simple first-angle projection of a building. The plan is drawn below and the elevations are drawn with reference to it, using the arrangement shown. The elevations are adjacent to one another.

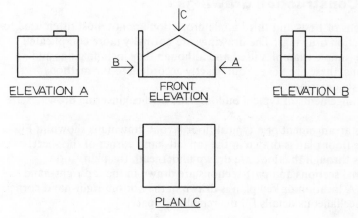

Fig. 3.3

The view looking left (A) is drawn on the left.
The view looking right (B) is drawn on the right.
The view looking down (C) is drawn underneath.

Third-angle projection

Figure 3.4 shows a simple third-angle projection of a building. The plan is drawn above and the elevations are drawn with reference to it, using the arrangement shown in the figure. The elevations are adjacent to one another.

The view looking left (A) is drawn on the right.
The view looking right (B) is drawn on the left.
The view looking down (C) is drawn above.

22

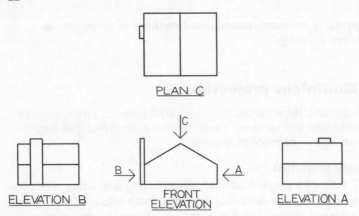

Fig. 3.4

3.5 Construction drawings

Variations of first- and third-angle projection are not most often used for construction drawings. The drawings are normally more complicated than the simple examples used and although they include plans and elevations, these are not set out exactly according to the methods previously described.

Arrangements of typical building and civil engineering drawings are shown.

The arrangement of a typical floor layout drawing is shown in Fig. 3.5. The floor plan is drawn in the top left-hand corner of the sheet. Sections through the floor are drawn underneath the plan, and additional sections and part-sections are drawn in the top right-hand corner. A location or key plan is drawn in the bottom right-hand corner and miscellaneous details fill the remaining space.

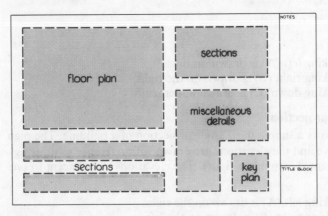

Fig. 3.5

Fitting the required information onto a drawing so that it is clear is more important than rigidly following the formal conventions of first- and third-angle drawing.

A typical layout of a general-arrangement drawing for a bridge is shown in Fig. 3.6. The positions of the elevation, section and plan are not fixed and the drawing layout may be modified. Often the size and shape of the views will affect the way in which a drawing is laid out.

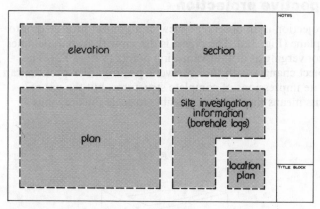

Fig. 3.6

3.6 Axonometric projection

There are three types of axonometric projection of which *isometric* is the most common.

An isometric drawing is a view of an object which has been rotated and tilted with respect to the plane of projection.

An isometric view of a rectangular block is shown in Fig. 3.7. The vertical axis of the block is drawn vertical. The other axes are drawn at 30° to the horizontal. The parallel sides of the block are not converging as they recede but are drawn parallel. Also all dimensions are drawn true to scale.

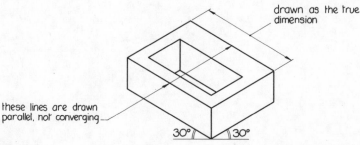

Fig. 3.7

An advantage of this type of three-dimensional projection is that it can be scaled, since everything is shown at its true size. However, although the drawing has a three-dimensional effect, it tends to look distorted because human perception expects objects to appear smaller as they recede. Isometric drawing can be very useful to clarify plans and elevations when a particularly complex detail has to be shown.

3.7 Perspective projection

Perspective projection is an observer's view of an object projected onto an imaginary plane (Fig. 3.8). The object is drawn with its parallel lines and surfaces converging at distant vanishing points (VP1, VP2). The view of the object changes with the observer's position. Perspectives can be used to create impressions of complete structures, but the use of vanishing points means that it is not possible to scale the drawings directly.

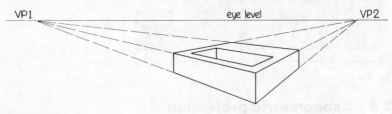

Fig. 3.8

Section 4

Linework and dimensioning

4.1 Introduction

This section describes the different types of linework used on construction drawings.

4.2 Line thickness

A typical drawing may include outlines, sections, hidden details and so on. Several different thicknesses of line should be used so that the finished drawing is well balanced and 'looks right'.

For the scales commonly used for structural drawings, BS1192: Part 1: 1984 British Standard for Construction drawing practice, recommends that the ratio of thickness between any two lines should be not less than two to one. Typical line thicknesses used on drawings are shown in Fig. 4.1.

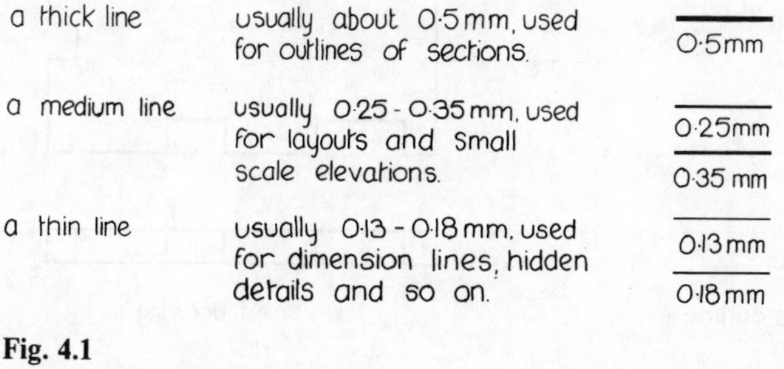

a thick line	usually about 0·5mm, used for outlines of sections.	0·5mm
a medium line	usually 0·25 - 0·35mm, used for layouts and small scale elevations.	0·25mm 0·35mm
a thin line	usually 0·13 - 0·18mm, used for dimension lines, hidden details and so on.	0·13mm 0·18mm

Fig. 4.1

The recommended thicknesses are suitable for drawings prepared on A2, A1, A0 and B1 drawing sheets, but they may not always be appropriate for A4- and A3-sized detail sketches. Individual drawing styles will also affect the appearance of a drawing, and it is not always necessary to adhere rigidly to the recommended thicknesses. The presentation and balance of a drawing are more important.

Other standard forms of line used on construction drawings are shown in Fig. 4.2.

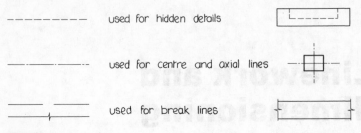

Fig. 4.2

4.3 Linework

The first stage in the preparation of a construction drawing is to draw the basic outlines of the plans, sections and elevations in pencil. The linework should be very fine and only just visible. This is best done with a very sharp, hard lead.

The pencil outlines are a framework which can be overdrawn in ink or softer pencil to produce a finished construction drawing.

The basic outline and finished drawing for the plan and elevation of a slab are shown in Fig. 4.3. If the pencil lines on the negative are very faint they will be invisible on a normal dyeline print and so the over-runs on the finished drawing need not be removed.

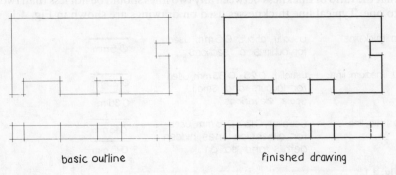

basic outline   finished drawing

Fig. 4.3

4.4 Cross-sections

When a cross-section is drawn it is normal to show the cut face shaded; this adds emphasis and gives depth to a drawing. For example, a reinforced concrete section (Fig. 4.4) may be: (a) shaded on the back of the negative using a coloured pencil (usually green), or (b) shown as the exposed face of the concrete. For very large sections it is not necessary to shade the whole of the face (c). A full range of shadings for different materials is given in section 6.

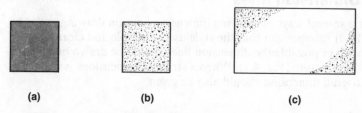

(a)   (b)   (c)

Fig. 4.4

4.5 Section and elevation markers

Four typical markers are shown in Fig. 4.5. BB shows a BS1192 recommended marker but the others are often used. The exact style is not important provided it is simple and clear. The markers must be located on the plane of the section or elevation with the indicators pointing in the direction of view (Fig. 4.6). While the sections markers are shown in the appropriate direction, the letters must be read from the bottom of the drawing (Fig. 4.7).

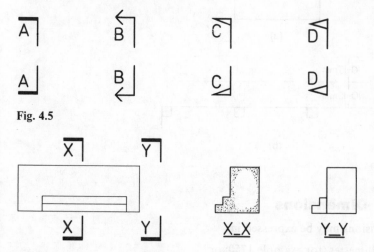

Fig. 4.5

Fig. 4.6

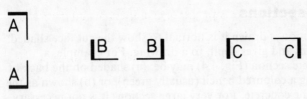

Fig. 4.7

4.6 Dimension lines

There are several ways of showing dimension lines on drawings
(Fig. 4.8). It is important that the style used is simple and clear.

Whenever possible the dimension lines should be drawn outside the
line of the structure (Fig. 4.9). When a string of dimensions is detailed
(b) the overall dimension should also be given.

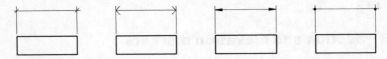

Fig. 4.8

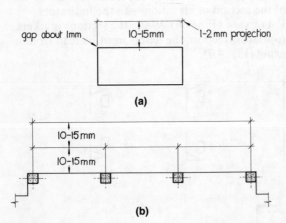

gap about 1mm   10-15mm   1-2 mm projection

(a)

10-15mm

10-15mm

(b)

Fig. 4.9

4.7 Dimensions

Dimensions may be expressed in:

(a) millimetres, for example 1750; or
(b) metres (to three decimal places), 1.750.

These two forms of presentation are visually compatible and there is no risk of misreading the dimension.

The symbols mm and m are not used.

Alternatively, certain dimensions may be written in an abbreviated form. For example 1700 may be shown as: 1.70 m or 1.7 m. The suffix m must be used for these.

When dimensions of 10 m and above are expressed in millimetres the 'thousands' marker should be shown as a gap, for example 17 500, 27 425. Dimensions should be written just above the line to which they relate, and whenever possible, should be positioned to read from the bottom or right-hand edge of the drawing (Fig. 4.10).

Occasionally a structure is angled in such a way that the associated dimensions cannot comply with this convention (Fig. 4.11). In such cases the dimensions should be shown in convenient positions. They must be clear.

When dimensions would be cramped, they should be drawn well clear of the object. Identifying arrows may be included (Fig. 4.12).

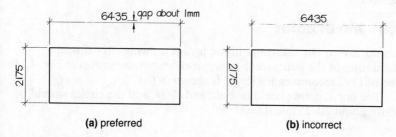

Fig. 4.10

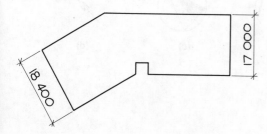

Fig. 4.11

4.8 Holes

Holes are shown as thin crossed lines (Fig. 4.13). Closed pockets are shown as a single diagonal line (b).

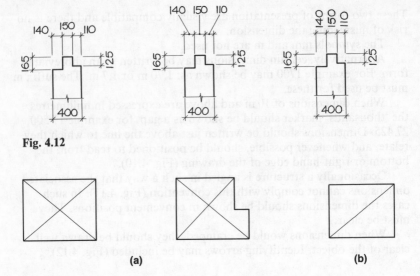

Fig. 4.12

Fig. 4.13

(a)   (b)

4.9 North point

It is normal to put a north point on layout drawings to define the orientation of the structure. Four typical designs are shown in Fig. 4.14. The BS1192 recommended shape is shown in (b).

The north point should be bold and clear, and the circles should be drawn about 20–25 mm diameter.

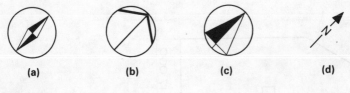

(a)   (b)   (c)   (d)

Fig. 4.14

4.10 Example

Figure 4.15 illustrates the use of linework on a typical construction drawing.

The drawing shows the general arrangement of a concrete roof slab for a pump chamber. When the chamber is completed, the holes will be fitted with steel access covers and a cabinet containing control gear will stand on the raised plinth. Bolts concreted into the pockets secure the plinth.

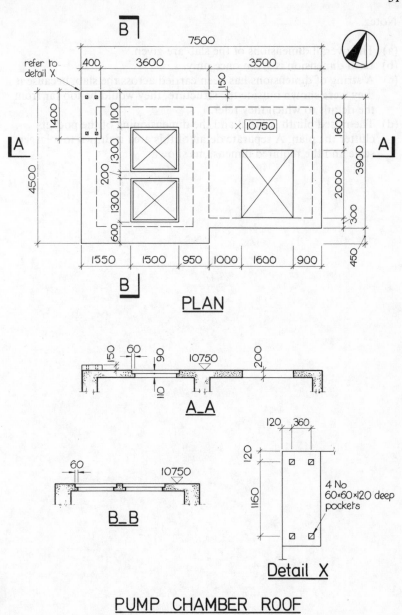

PLAN

A_A

B_B

Detail X

4 No
60×60×120 deep
pockets

PUMP CHAMBER ROOF

Fig. 4.15

32

Note:

(a) The overall dimensions of the slab are given.
(b) Each dimension is given once only.
(c) A string of dimensions has been carried across the slab, because if they were shown outside the structure, they would be too far from the details to which they refer.
(d) The raised plinth is small and the dimensioning of the pockets clutter the plan. A separate detail of this area is drawn to a larger scale and the required dimensions shown.

Section 5

Setting out structures

5.1 Introduction

When a set of construction drawings is prepared, it is essential that
details are included which enable the structure to be located accurately
on the site. This section describes the basic setting-out information used
for structures.

5.2 Grid referencing

1:50 000 Ordnance Survey (OS) maps are set out on a 2 cm square grid.
At that scale 2 cm represents 1 km. Figure 5.1 shows a detail from a
1:50 000 OS map. The grid lines are numbered. The north–south running
lines, 358000, etc. are called *eastings*, and the east–west lines, 173000,
etc., are called *northings*. These are the distances in metres east and
north of a datum point called the 'false origin', which is approximately
100 km west of the Scilly Isles.

Any point on the map can be located by a grid reference (or
coordinates). Thus the location for point X would be E356500, N173400.

The system is called the National Grid.

5.3 Surveys

Larger-scale surveys are required to position a structure. These are
normally 1:500 or 1:200. When land surveyors prepare an enlarged

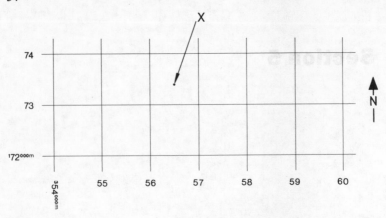

Fig. 5.1

survey of a particular area, they set up an independent baseline which
can be easily identified by road nails, steel pins concreted into the
ground, or from existing features such as manhole covers. This line can
be used later to establish the coordinates for the setting out of the
structure.

On large sites, such as roadworks projects which may be several
kilometres long, an independent local grid may be established. This can
be referenced back to the National Grid.

5.4 Measuring height

The height of the ground at any location can be measured from a datum
point called a *bench mark*. Bench marks are local stations established by
the Ordnance Survey which are referenced back to a zero datum plane
taken to be the mean sea-level at Newlyn in Cornwall. This is commonly
referred to as the *ordnance datum* (OD). The height above the sea-level
datum is called the *level* and is measured in metres. An Ordnance Survey
bench mark is referred to as an OBM. When a structure is to be built, a
temporary bench mark (TBM) is set up on the site. It may be a steel pin
concreted into the ground, or an existing feature such as a manhole
cover. The level of the TBM is established from the nearest OBM.

Individual levels required in the structure, such as the level of a
foundation soffit or a floor slab, are taken from the TBM. Vertical
dimensions in a structure are identified by levels (OD). They are the
common referencing system for all aspects of a project. For example,
drainage is laid to levels. A large-diameter drain which crosses the line
of a building may have severe implications to the layout of the
foundations, and the design of suitable details starts with a comparison
of levels.

5.5 Setting out buildings

The setting out of a building requires the establishment of base lines for the location of the whole building on its site, and the setting up of a grid system to locate individual elements within the structure.

On a green-field site where there are no existing buildings, the base lines are normally set out on the axes of the new structure (Fig. 5.2). The building baselines are referenced back to setting-out points 1, 2, 3 and 4 which may be steel pins concreted into the ground. The coordinates of these points are established from the independent base line which is usually set up during the initial survey. An existing manhole cover which is not to be disturbed during the new construction may be used for the TBM.

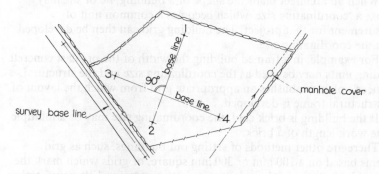

Fig. 5.2

It is not essential to have two baselines at right angles as shown in the figure: alternative layouts may be used.

When a new building is sited in a built-up area, for example a city centre, the baselines chosen may be the lines of an adjacent building or a kerb line.

5.6 Building grids

The plan shapes of buildings are often rectangles or a series of linked rectangles (Fig. 5.3(a)). The system normally used to locate the individual elements within such a building is to set up vertical and horizontal lines on the plan shape of the structure. This is the building grid. The grid is identified by either letters across and numbers down (b) or vice versa.

The building grid is usually planned to pass through commonly occurring elements in the structure, such as columns. It is not necessary for the vertical and horizontal grids to have identical spacings, which would give a square layout, but it is preferable that the spacing is constant between grid lines in the same row.

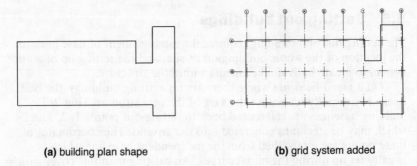

(a) building plan shape   **(b)** grid system added

Fig. 5.3

When an architect plans the shape of a building, he or she may choose a 'coordinating size' which becomes a common unit of measurement for that project. The building grid can then be developed from the coordinating size.

For example, in a framed building the width of the pre-cast concrete cladding units may be used as the coordinating size and the structural engineer's plan establishes an appropriate grid from which the layout of the structural frame is developed.

If the building is brick clad, the coordinating size may be a multiple of the work length of a brick.

There are other methods of setting out buildings, such as grid systems based on a 100 mm or 300 mm square, or grids which mark the external boundaries of the columns and other structural elements, rather than passing through their centre lines.

5.7 Referencing

The grid system is only used as a means of referencing the structure and it is not necessary for all the elements to be designed to fit onto the particular layout chosen. When, for example, a column does not come on the established grid, it is simply referred to as 'off-set' and its position established as shown in Fig. 5.4. Note that the grid has been drawn only where it passes through the structure.

5.8 Working with grids

The grids act as a common referencing system between architect, engineer and contractor and have the obvious benefit that they allow all parts of the structure to be easily identified. This is particularly useful in telephoned or written communication.

Also, during the early stages of a design, when the shape of a

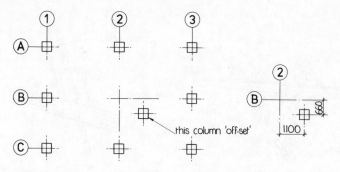

Fig. 5.4

building is being developed, it is very helpful to be able to move the elements of a structure on a fixed grid.

At the start of construction, the contractor sets out the baselines and the building grid.

The construction drawings must show full setting-out details, including the position of the grid relative to the baselines, and the location and value of a TBM. The information may be prepared on a special setting-out drawing as shown in Fig. 5.5, or included on one of the construction drawings, for example the foundation layout.

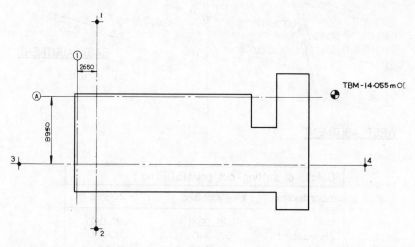

Fig. 5.5

5.9 Tanks and reservoirs

Concrete tanks and reservoirs are often rectangular in plan and the techniques used for setting out buildings with baselines and grid systems can also be applied to them.

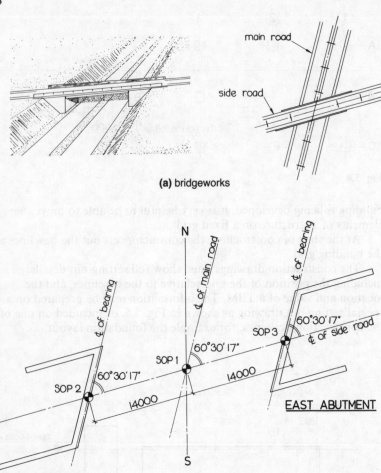

(a) bridgeworks

Details of setting-out point (SOP) no.1		
coordinates	X = 4938·325	Y = 2386·675
	main road	side road
chainage (m)	1675·240	280·600
bearing	14° 38′ 25″	75° 08′ 42″

(b) setting-out details

Fig. 5.6

5.10 Setting out bridges

A highway bridge is normally positioned from the setting out of the road which it crosses or carries.

Roads are usually designed with the aid of computers using highway design programs, which make the complex calculations necessary in determining the carriageway alignment. A specific distance measured along the road is called the *chainage*. Chainage is measured in metres from a datum point which may be one end of a road scheme. A typical computer output gives the coordinates and the bearing of the road at intervals of chainage (bearing is effectively the direction of the road measured as an angle from the north/south axis).

Typical setting-out details for a bridge are shown in Fig. 5.6. The bridge carries a side road over a main road (a). The setting-out drawing (b) shows the centre-line axes of the roads and a simplified outline of the abutments. Coordinates and bearings taken from the road design are used to establish the location of the bridge. The intersection of the centre-line axes is taken as the setting-out point SOP 1. Dimensions are given from SOP 1 to the centre line of the bridge bearings. They are the principal dimensions linking the deck and abutments. The intersections of the road and bridge bearing axes are identified as SOP 2 and SOP 3. These dimensions and setting-out points are also shown on the abutment layout drawing where the bearing centre lines become the reference axes from which the abutments are set out.

The coordinates used for the setting out of this bridge are based on a local grid.

An alternative method for setting out bridges is to tabulate X and Y coordinates for the principal points on the structure.

A bridge at right angles to the road crossed is commonly referred to as a *right bridge* (Fig. 5.7(a)). A bridge not at right angles is described as *skew* and the angle of skew may be include on the setting-out details (b).

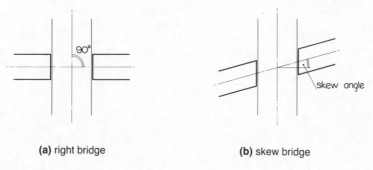

(a) right bridge   **(b)** skew bridge

Fig. 5.7

5.11 Drawing grids and baselines

The grid and baselines should be drawn very fine—0.13 or 0.18 mm pen or equivalent.

It is quite common for completed drawings to require modification. Rubbing or scratching out the lines causes the surface of the drawing sheet to become roughened and, as a result, new lines drawn on this surface are often blurred.

If, when the drawing is being prepared, the negative is turned over and the grid lines are inked on the back of the sheet, they will then be left undisturbed if alterations are made.

Section 6

Construction details

6.1 Introduction

This section considers aspects of construction which influence the design and detailing of structures.

6.2 Building structures

The design of a building is normally carried out by independent teams of architects, structural engineers, mechanical and electrical consultants and quantity surveyors who collaborate as a project team to produce the complete design.

Work starts with preliminary schemes which are used to establish the basic shape and form of the building, and continues with detailed design calculations and the working drawings required by the contractor to carry out the construction.

Although the members of the project team have a common purpose in designing the building, each is responsible for a different aspect and sees the project from his or her own point of view. The architects, as the principal consultants, are responsible for the function and appearance of the completed building. The engineers design the foundations and structural framework, the mechanical and electrical consultants are concerned with the heating, electrical and other services and the quantity surveyors prepare estimates and cost comparisons as the building is designed and advise on the economics of systems, materials and so on.

A cost breakdown for a typical reinforced-concrete-framed building is approximately:

architectural — 40%
structural — 20%
services — 40%

which means that for each £1m. of contract value, about £200 000 is for the structure. For an extensively serviced building such as a hospital, the service element increases, and for buildings with unusual or complicated foundations the percentage cost of the structure may rise.

The structural design team prepare drawings of the foundations and the building frame. The drawings also show details such as the layout of bolt and dowel fixings and the holes required by the mechanical and electrical consultant for major heating and ventilation services. Smaller services such as water pipes and electrical trunking are usually attached with simple plug-and-screw fixings and do not affect the structure.

6.3 Freehand sketching

Freehand sketching is an important aspect of structural detailing enabling the design team quickly to develop the details in a structure. Often a set of sketches is used as the basis for a working drawing. Much of the information used is abstracted from architects' and mechanical and electrical consultants' drawings and may require interpretation before being used by the design team. Used for this purpose the freehand sketch becomes an intermediate stage in the preparation of the structural drawings.

Sketching is also used on site to record the results of structural surveys and trial-pit site investigations, and to plot the positions of water, gas and other services exposed as part of the exploratory excavations. Although the locations of public services are charted, excavations to confirm their positions are essential when new works may be built close by.

Sketches are best done in pencil so that changes are easily made as the details develop, and although they should be drawn roughly to scale, exact drawing is not normally necessary since the main purpose of the sketch is simply to record all the information required for the working drawings, which are then drawn to scale. Examples of typical sketches which may be drawn by the design team are shown in Fig. 6.1.

When pre-cast concrete cladding panels are used on a building they normally stand on the edge of the reinforced concrete floor slab and are attached to the building frame with mechanical fixings, usually dowels or bolted cleats. Much of the design of the panels, the appearance, glazing, waterproofing details and so on, would be the architect's responsibility, but the structural design team normally advise on suitable fixings and incorporate them in the structure.

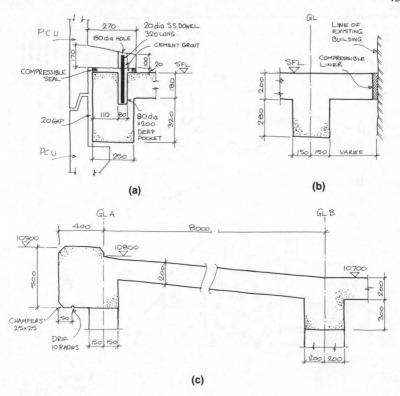

Fig. 6.1

The freehand sketch (Fig. 6.1(a)) shows the arrangement of pre-cast units (PCUs) on the slab edge. The detail is used to establish the basic geometry of the joint, including the depth and diameter of the pocket and the edge distances for the slab and the nib on the pre-cast unit. Details of the dowel, grout and seal are also shown.

The information shown on this sketch will be used on the structural layout and reinforcement drawings and on the drawings of the pre-cast concrete cladding panels.

The sketch (b) shows a detail that is often used when a new structure is built against an old building which has substantial brick or masonry walls of varying thickness and alignment. The structural frame is detailed to include an infilling cantilever slab which is cast against the old wall. The compressible liner prevents the interlocking which would occur if the concrete were cast directly against the rough surface. The information on this sketch would be used on all appropriate floor slab drawings for the building. The span of the cantilever varies and this should be noted on the structural drawings. A special reinforcement detail would be required.

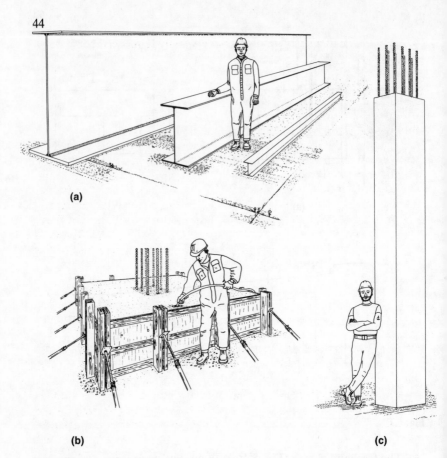

(a)

(b)

(c)

Fig. 6.2

Freehand sketching can also be used to sort out important details in a way that would be difficult on the structural drawings.

In construction, vertical distances are normally considered in terms of their relative (ordnance datum) levels. The sketch (c) is used to examine the relationship between adjacent beam details forming part of an exposed concrete floor slab. For drainage, the slab is detailed to a fall, and although the change in levels is 100 mm, over 8 metres this would be a slope of less than 1°. Drawn to scale this would be difficult to appreciate. By sketching the separate beam details much closer together than they will be in the structure, the vertical relationship between the two is immediately apparent.

Even good freehand sketches can be drawn only roughly to scale and occasionally it is necessary to prepare a simple scale drawing to check the layout and proportions of part of a structure. The drawing might be used to determine the levels of foundations on a sloping site or to examine the arrangement of bars in a heavily reinforced concrete structure.

6.4 Structural sizes

Building and civil engineering works are large in human terms. It is important for the design team to appreciate the eventual size of the structures being drawn and to anticipate the construction problems that may have to be overcome. The 'on-site' sketches (Fig. 6.2) indicate the size of different structures. Typical structural steel I-sections are shown (a). The smaller ones represent the smallest and largest of a range of universal beams which are rolled from steel billet to standard profiles. Universal beams are used extensively in structural steel frameworks. The large I-section is typical of a range of welded plate girders that are made from separate steel plates continuously welded to form the I-shape. Sections of this type can be made almost 4 m deep and are used in bridges and heavy industrial structures.

The pad foundation (b) will support a single column in a tall building. It is about 3 metres square (the size of a small room) and almost a metre thick. Steel reinforcement for the column will be connected to the bars which project from the foundation (starter bars).

Usually, at the beginning of a contract, a major excavation is carried out for the construction of the foundations, and the building site quickly becomes a very large hole in the ground. For even a moderate sized building, several thousand cubic metres of soil may be removed for the construction of foundations and basements. Awareness of these major works is quickly lost because the completed foundations are buried and a ground floor slab cast before work on the superstructure begins.

The reinforced concrete column (c) is about 5 metres high and might be used in the entrance area of a large building—a hospital, airport terminal, offices, etc. The concrete is cast in a timber mould (shutter). The practical limitations in placing and adequately compacting concrete mean that tall slender elements are often cast in several stages (lifts). The column shown is approaching the limit of construction as a single pour, and although casting is the contractor's responsibility, the design team must be aware of the practical problems involved and prepare details which allow for construction difficulties. For a taller column, where it is clear that the concrete should be cast in several lifts, the design team must prepare suitable details allowing for joints in the reinforcement and the concrete.

6.5 Estimating size

Because the design team draw very large structures to very small scales, it is important to be able to estimate and appreciate the actual size of

46

proposed details, and there are many simple ways of doing this. For example, knowing the distance between columns in a large office or the approximate width and height of surrounding buildings may give useful guidance in judging dimensions on a current project. Markers placed at metre centres along the office wall can help the design team visualise lengths in the 3–10 metre range, and a 3-metre steel tape is useful in appreciating the actual size of pockets and holes in slabs, and the cross-sections of beams and columns.

Many manufacturers of components used in construction supply samples of their products, and design teams often acquire a collection of ties, bolts, waterproofing materials and so on. Seeing and handling construction materials gives valuable guidance in preparing structural details. A visit to the site of proposed works and site photographs also help the design team.

Section 7

Project drawings

7.1 Introduction

The structural drawings prepared by the design team may form part of the contract which is made by the engineer or architect (on behalf of the client) and a contractor. They are in effect legal documents. The contractor undertakes to build what is shown on the drawings, which should be clear and unambiguous. As a contract proceeds, areas of uncertainty on drawings quickly become apparent and may be the subject of contractors' claims for additional payments or extensions of time.

The drawings are read in conjunction with contract documents which contain details of the contract procedures, specifications of materials and workmanship and bills of quantities for the works. Different forms of contract are used for building and civil engineering projects.

This section describes typical drawings required for building and civil engineering structures and explains how they relate to the contract. The scales indicated in the description of each drawing are typical.

7.2 A reinforced-concrete-framed building

For buildings such as hospitals, schools and office blocks an architect is the principal consultant and coordinates a project team which includes a structural engineer who is responsible for the design and detailing of the

48

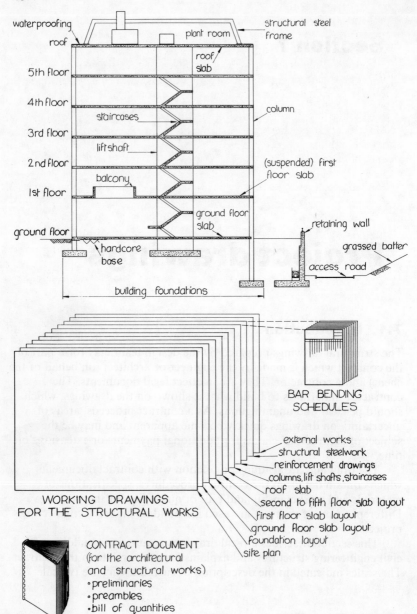

Fig. 7.1

structural works, a mechanical and electrical consultant who is responsible for the building services and a quantity surveyor who prepares the bills of quantities and monitors costs.

Figure 7.1 shows the layout of a conventional multi-storey office block. The structure is reinforced concrete. A typical set of working drawings prepared by the structural design team are:

Site plan: the site plan is normally drawn from the ground survey and is used to locate the baselines from which the building is set out, and to show the building grid.

Foundation layout: the foundation layout shows the plan, elevations and sections of the building foundations (scales 1:200, 1:100, 1:50). Notes box entries include items for the concrete quality and safe ground bearing capacity required at foundation level. Allowable bearing capacities are calculated from the site investigation data. When the foundations are dug, the ground may not be exactly as indicated by the site investigation, requiring the possible removal of unsuitable material and its replacement with mass concrete.

Ground floor slab layout: the ground floor slab is normally cast directly onto a compacted hardcore sub-base, and although the drawing basically illustrates a simple slab of constant thickness (usually 150–200 mm), there are always a number of details to be shown, such as door thresholds, columns, lift shafts and staircases. Manholes and ducts may also be required for the services coming into and out of the building.

The ground floor slab layout is usually the first drawing on which the plan form of the building superstructure becomes apparent.

First floor slab layout: the entry levels into a building often include particular architectural features such as assembly areas and balconies which are not repeated on the floors above. Special structural details may be required.

For the building shown, a separate drawing is needed for the slab at first-floor level. The drawing includes a plan and sections. The sections are chosen to pick up details such as slab edges and upstands which are not completely defined on the plan.

Second to fifth floor: in tall symmetrical buildings, the upper floors are normally structurally similar and one slab drawing may serve several floors. The content is similar to the first-floor layout.

Roof slab: the lift motors, air conditioning and other mechanical equipment required to service the building is normally housed in plant rooms on the roof. Exposed areas around the plant rooms are fully waterproofed. Numerous holes, pockets, plinths and upstands are required to accommodate the plant and waterproofing and a special roof slab drawing is prepared.

Columns, lift shafts and staircases: separate drawings are required for the columns, lift shafts and staircases. These elements are normally shown as a series of plans and elevations with the floor slab at each level shown in section. The liftshaft drawings usually include details of proprietary cast-in fixings required for the lift installation. The stairs are normally finished with screed and the structural profiles require careful detailing.

Reinforcement drawings: a set of drawings are prepared which show the type, size and layout of all the reinforcing bars required in the concrete structure. It is usual, though not essential, to match the reinforcement drawings with particular concrete profile drawings so that for each element in the structure, foundations, floor slabs, etc., there are complementary profile and reinforcement drawings.

The reinforcing bars are pre-bent to a range of standard shapes and given identification numbers (bar marks). The drawings are read in conjunction with (A4-size) bar schedules which list the bars by mark number and give full details of the shapes and dimensions of each. Notes box entries are required for the protective concrete cover to the reinforcement, construction joints and bar schedule references.

Structural steelwork: reinforced-concrete building structures often incorporate some structural steelwork. In the building shown, the plant room is a steel frame with profiled sheet metal cladding and there may be also be some steelwork at the lower levels to support entrance canopies and other architectural features. Separate steelwork drawings are prepared for these elements.

External works: there are often external works adjacent to large buildings, such as buried chambers for tanks, ramps and retaining walls. They are shown on external works drawings.

Contract document

Building works are normally carried out using a Joint Contracts Tribunal (JCT) form of contract. There are several versions appropriate to different types and sizes of project and they cover the architectural and structural engineering aspects of the work. Mechanical and electrical services are usually dealt with in a sub-contract which is let independently. The whole of the work is under the supervision of the architect.

The JCT contract documents for the building shown would comprise general conditions and preliminaries, preambles, bills of quantities and contract drawings. Details of the content and presentation of the document may vary, but in general:

General conditions and preliminaries: in keeping with many legal documents, the contract starts by naming the various parties involved: the employer (client), the architect, the structural engineer, the mechanical and electrical consultant and the quantity surveyor. This is followed by a brief description of the site, of the works to be undertaken and a list of the contract drawings which may include tender drawings prepared by the structural design team.

The conditions of contract follow. They cover all aspects of the running of the contract, including the setting out of the works (for which the contractor is responsible), temporary works, testing of materials, management of the site, safety, insurance, methods of payment and itemised provisional sums. These conditions are often only a list of numbers of clauses in the standard form of contract.

Preambles: the preambles include descriptions of the materials and workmanship required for the architectural and structural aspects of the works. Engineers often have their own 'in-house' specifications for the structural works which are incorporated into this section.

Bill of quantities: the bill of quantities lists the items to be priced in carrying out the works. Each item is described and quantified in the document and is priced by contractors tendering for work.

7.3 An example of a reinforced-concrete pedestrian subway under a highway

For the design of highway structures, the engineer is the principal and usually the only consultant. A major highways scheme usually includes roadworks and several structures such as bridges, culverts, retaining walls and pedestrian subways. Figure 7.2 shows a reinforced-concrete pedestrian subway on piled foundations. In addition to designing and detailing the structure the design team is normally responsible for architectural and services aspects of the work. For a large structure an architect may be employed to advise on the visual aspects. Typical drawings required for the subway are:

Cover sheet: this is used as the title sheet, giving the name and classification of the road and the name of the scheme, the client and the designers. Large dry transfer letters are normally used.

List of drawings: numbers and titles of the drawings forming the set.

General arrangement: this is a basic drawing of the complete structure, showing a plan, elevations and typical sections (scale 1:50), a site location plan (scale 1:2500) and the borehole records taken from the site investigation report. Notes box entries include a list of drawings for the structure. The main purpose of the general arrangement (GA) is to give an immediate impression of the scope of the works, and it is the only drawing which shows the whole structure assembled. Other drawings usually show components of the structure.

Piling details: this drawing shows a setting-out plan (scale 1:50) for the pile positions with individual piles numbered for identification, details of an individual pile showing the profile and reinforcement details, tables listing natural ground levels, pile cut-off levels, expected length and axial loads and bending moments imposed by the structure. It is usual to load-test one or more piles. Full details are given, including a table showing the increments of loading to be applied.

Concrete profile drawings: drawings for the *in-situ* works, showing plans, elevations and sections of the barrel floor, walls and roof, the head and

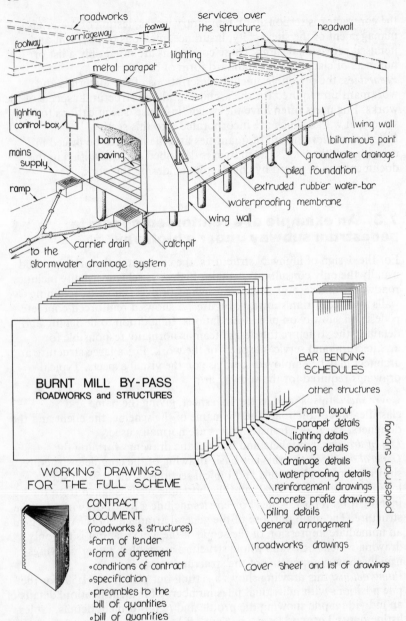

roadworks
carriageway
footway
footway
services over
the structure
headwall
lighting
metal parapet
lighting
control-box
barrel
paving
mains
supply
ramp
wing wall
bituminous paint
groundwater drainage
piled foundation
extruded rubber water-bar
waterproofing membrane
wing wall
carrier drain
catchpit
to the
stormwater drainage system

BAR BENDING
SCHEDULES

BURNT MILL BY-PASS
ROADWORKS and STRUCTURES

other structures
ramp layout
parapet details
lighting details
paving details
drainage details
waterproofing details
reinforcement drawings
concrete profile drawings
piling details
general arrangement

pedestrian subway

roadworks drawings
cover sheet and list of drawings

WORKING DRAWINGS
FOR THE FULL SCHEME

CONTRACT
DOCUMENT
(roadworks & structures)
∘form of tender
∘form of agreement
∘conditions of contract
∘specification
∘preambles to the
bill of quantities
∘bill of quantities

Fig. 7.2

wing walls (scale 1:50). Also larger-scale (1:20, 1:10) details of service entries, drainage channels and exposed concrete finishes and notes box items for the quality of concrete, surface finishes and construction joints.

Reinforcement drawings: these drawings show details of the type, diameter and location of the reinforcing bars required in the *in-situ* concrete structure. Bars are pre-bent to a range of standard shapes and given identification numbers (bar marks). The drawings are read in conjunction with (A4-sized) bar schedules which list the bars by mark number and give full details of the shapes and dimensions of each. Notes box entries are required for the protective concrete cover to the reinforcement, construction joints and bar schedule references.

Waterproofing details: the construction joints in the concrete structure are protected by extruded rubber water-bars, a waterproofing membrane is fixed externally to the barrel floor, walls and roof, and the backs of the wing walls below ground level are finished with bituminous paint. The waterproofing is used to protect the concrete structure from the corrosive effects of road salts and continuous exposure to groundwater.

Drainage details: most highway structures need sufficient storm and groundwater drainage to require a separate drainage drawing showing filter materials, pipework rodding eyes, catchpits and connections to the main stormwater drainage system.

Paving details: the paved areas inside the structure are laid to falls so that wind-blown rain can drain away. The paving drawing shows details of the paving materials—slabs on lean-concrete, asphalt, etc., and the levels.

Lighting details: subways are normally lit by encased striplights which are set in pre-formed recesses in the roof or cornice. This drawing shows details of the proprietary lighting units and associated wiring, fuseboxes, control gear and the connection to the mains supply.

Parapet details: the metal parapet is normally designed around one of a number of proprietary systems, and the design team's drawing is prepared using details of standard components. The parapet is made by a specialist fabricator who uses these details to prepare workshop drawings.

Ramp layout: the carriageway over the structure is detailed with the roadworks drawings. The pedestrian ramps into the subway and the associated earthworks may be detailed with the structure or with the roadworks.

Contract document

Highway works are normally carried out using an Institution of Civil Engineers (ICE) form of contract—a 'civils' contract. The contract document covers both roadworks and structures, and a typical document contains:

Form of tender: a pre-written letter which has to be signed by the contractor and constitutes an offer to carry out the works.

Form of agreement: a copy of the document which will later be signed by the client and the contractor, confirming the contract.

Conditions of contract: the conditions of contract are based on a standard document prepared by the ICE and others, which covers all aspects of the organisation and day-to-day running of the contract, including the provision of adequate temporary works, site supervision, insurance, payment of monies to the contractor and dealing with claims for unforeseen conditions and additional works.

Specification: the specification used for highway works is the 'Specification for Highway Works' (formerly the Specification for Road and Bridgeworks) prepared by the Department of Transport and others, published by Her Majesty's Stationery Office. This specification is referred to in the contract document, but is not reproduced.

Preamble and Bill of quantities: work for highway projects is measured and itemised in accordance with the standard 'Method of Measurement for Road and Bridgeworks', prepared by the Department of Transport and published by HMSO.

It is particularly important that the wording used in the specification and method of measurement is also used on the drawings. In this way there should be no possibility of misinterpretation, which could lead to expensive claims by the contractor.

Copies of the bar schedules may also be incorporated in the contract document as an appendix.

7.4 Design variations

It may not always be possible to produce exact, detailed drawings of every part of a structure. For example, the design and detailing of foundations is based on the findings of a site investigation carried out by boring holes or excavating at isolated locations. Only during construction, when the mass excavation for the foundations is carried out, can the assumed ground conditions be confirmed across the site.

Possible variations are considered at the design stage and the drawings must include appropriate details. Provisional items are included in the bill of quantities and are priced by the contractor. This establishes a rate for the works should they become necessary.

A similar situation can occur when an old building is redeveloped by demolishing the original interior, leaving the external shell and then building a steel or concrete frame on the site anchoring the original shell to the new structure. An extensive structural survey is normally undertaken at an early stage of the design, but this cannot guarantee to identify every construction problem. The design team must anticipate this by preparing details which can be adjusted to accommodate possible variations.

Section 8

Symbols and abbreviations

8.1 Introduction

Symbols and abbreviations are used extensively on construction drawings. Many are listed in British Standards—principally BS1192, others are tabulated in technical standards which deal with specific aspects of construction or are otherwise in general use. This section lists commonly used symbols and abbreviations.

8.2 Symbols for cross-sections

Parts of a structure shown in cross-section are emphasised using symbols, or shading, which also adds perspective to a drawing. Figure 8.1 shows suitable symbols and shading.

8.3 Abbreviations

Abbreviations for components and materials
Abbreviations used for components and materials commonly specified on structural drawings are shown in Fig. 8.2.

General abbreviations for drawings
Abbreviations and symbols used for dimensions, levels and other aspects of draughtsmanship are shown in Fig. 8.3.

56

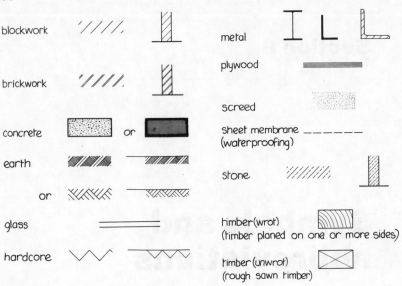

blockwork		metal
brickwork		plywood
concrete	or	screed
earth		sheet membrane (waterproofing)
or		stone
glass		timber(wrot) (timber planed on one or more sides)
hardcore		timber (unwrot) (rough sawn timber)

Fig. 8.1

aggregate	agg	granolithic	grano
aluminium	al	hardcore	hc
asphalt	asph	hardwood	hwd
bitumen	bit	inspection chamber	ic
blockwork	blk	joist	jst
brickwork	bwk	material	matl
building	bldg	pitch fibre (pipe)	pf
cast iron	ci	plate	plt
column	col	rainwater pipe	rwp
concrete	conc	reinforced concrete	rc
damp-proof course	dpc	rodding eye	re
damp-proof membrane	dpm	stainless steel	ss
foul sewer	fs	storm water sewer	sws
foundation	fdn	tongued and grooved	t&g
glazed pipe	gp	(boarding)	

Fig. 8.2

approximately	approx	invert level (drainage)	i l
average	av	left hand	l h
bench mark (ordnance survey)	o.b.m.	long	lg
		maximum	max
bench mark (temporary)	t b m	metre	m
centres	crs	millimetre	mm
centre line	₵	minimum	min
centre to centre	%c	north point (typical)	
chamfered	cham	not to scale	nts
cover level (drainage)	c l	number	no
diameter (in a note)	dia	outside diameter	o d
diameter (before a dimension)	∅	radius (in a note)	rad
drawing	dwg	required level (on a plan)	× ‾10000‾
existing level (on a plan)	× 10000	required level (on a section)	‾10000‾ ▽
existing level (on a section)	10000 ▽	right hand	r h
existing ground level (in a note)	egl	rise of a ramp	1:15 →
external	ext	rise of a stair	‖‖‖→‖
figure	fig		
finished floor level	ffl	sheet	sh
hectare	ha	sketch	sk
holes	hls	specification	spec
horizontal	hor	square	sq
inside diameter	id	structural floor level	sfl
internal	int	vertical	vert
invert (drainage)	inv		

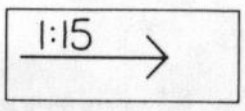

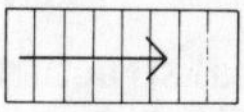

Fig. 8.3

Specific abbreviations

The final tables show abbreviations used specifically on reinforced
concrete detail drawings (Fig. 8.4), and structural steelwork drawings
(Fig. 8.5).

58

bottom	B	alternately placed	AP
top	T	alternate bars reversed	ABR
near face	NF	reinforcement	
far face	FF	(by type and diameter)	
each face	EF	high yield bars	T10, T12 etc
each way	EW	mild steel bars	R10, R12 etc

Fig. 8.4

universal beam	UB	rectangular hollow section	RHS
universal column	UC	square hollow section	SHS
rolled steel joist	RSJ or JOIST	butt weld	BW
rolled steel channel	RSC or ⊏	fillet weld	FW
angle	ANGLE or ∠	countersunk	CSK
tee	TEE or T	hexagon head	HEX HD
circular hollow section	CHS	mild steel	MS

Fig. 8.5

8.4 Example

Typical structural details using symbols and abbreviations are shown in Fig. 8.6. It is not always essential to use the abbreviated forms. For a single detail, material descriptions and other terms may be fully written (a), but when many similar details are shown on a drawing, abbreviations are usually more appropriate (b).

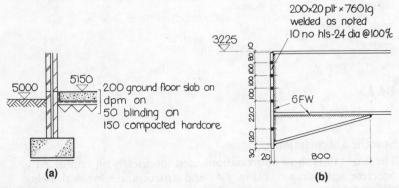

Fig. 8.6

8.5   Use of capital or lower-case letters

Both capital and lower-case letters have been used in the tables. Certain abbreviations are always shown as capitals. These are steel reinforcement R10, T10, etc., shown in Fig. 8.4 and the steel sections UB, UC, RHS, etc., shown in Fig. 8.5. Otherwise the choice usually depends on what is most appropriate in a particular situation. Generally, abbreviations written in capital letters appear more imperative, and are particularly suited to isolated references such as TBM, SFL and so on; whereas lower-case letters are more appropriate to the descriptive notes (eg. 25×25 cham $\times 1750$ lg), or the text written on the notes box. The choice may also be influenced by individual drawing styles.

Whatever form is used, abbreviations must be clear, consistent and unambiguous.

Section 9

Trade literature

9.1 Introduction

Proprietary products are used extensively in construction and items such as ties, inserts, concrete additives and waterproofing, as well as complete floor and roofing systems are regularly specified on structural drawings.

The design team use trade literature published by the manufacturers which contains details of their product ranges and other technical information. Where appropriate, strengths and safe loads are tabulated and clear dimensioned details of the products are included. These are needed to prepare the details on the drawings. In addition, most companies run a technical advisory service to answer questions about the application of their products.

Examples are shown for the use of manufacturers' trade literature for a flooring system and waterproofing details.

9.2 Concrete flooring

The first example shows the use of trade literature supplied by the manufacturer of a proprietary concrete floor system. The flooring is in the form of hollow pre-cast concrete planks which are factory made, transported to site and lifted directly into place onto supporting walls (Fig. 9.1). The structural floor is completed with *in-situ* concrete which fills the gaps between the individual units. The floor may be completed with a sand : cement screed which gives a uniform surface

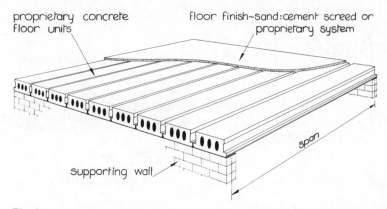

Fig. 9.1

to receive thermoplastic tiles, etc. This type of flooring system is often used in low- and medium-rise accommodation blocks, hostels, flats, commercial developments and multi-storey frameworks.

The extract from the trade literature (Fig. 9.2) shows cross-sections through the floor units and the tables which are used to select a suitable depth of floor unit for a particular combination of loads and spans. In the example assume that the floor spans 6 m and carries a distributed load of 5 kN/m² exclusive of self-weight and finishes. Referring to the table, a 150 mm deep unit is required. The maximum allowable span for these units is 6.35 m, but the actual load/span values rarely match the allowable figures exactly. The unit thickness should always be chosen to err on the safe side.

On the drawings, the units are called up using the manufacturers'

Pre-cast Concrete Floors

Overall Structural Depth mm	Dead Load 400 wide kN/m²	1200 wide kN/m²	Distributed loads in kN/m² Includes self weight and finishes 1.5kN/m²								
			0.75	1.5	2.0	2.5	3.0	4.0	5.0	10.0	15.0
			Maximum span in metres								
110	2.16	2.16	5.0	5.0	5.0	5.0	5.0	5.0	4.88	3.84	3.27
150	2.53	2.32	7.0	7.0	7.0	7.0	7.0	6.77	6.35	5.02	4.28
200	3.26	2.94	9.05	9.05	9.05	9.05	8.72	8.14	7.67	6.14	5.28
250	3.92	3.49	11.25	11.23	10.78	10.38	10.02	9.41	8.89	7.18	6.18
300	-	3.95	13.8	13.7	13.3	12.9	12.5	11.8	11.2	9.1	7.9

Fig. 9.2

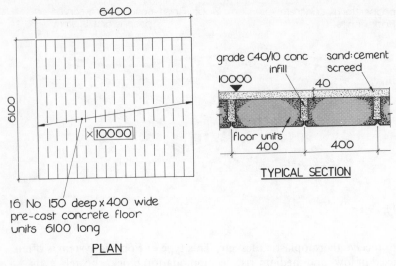

16 No 150 deep x 400 wide
pre-cast concrete floor
units 6100 long

PLAN

Fig. 9.3

name and references. The units can be supplied to the length detailed on
the drawings (Fig. 9.3). The trade literature also includes typical details
showing recommended methods of forming holes in the floors, slab edge
treatments, etc.

9.3 Waterproofing systems

Many types of reinforced-concrete structure require waterproofing:
basements in buildings, tanks, retaining walls and subways may all
require protection. Specialist manufacturers produce a range of
waterproofing products such as ribbed PVC water-bars, self-adhesive
sheet membranes, sealing compounds, fillers, liners and so on. The
second example shows waterproofing materials used at a joint in a
retaining wall. External retaining walls expand and contract due to
temperature changes and to accommodate these movements a small gap
(usually 20 mm) is required in the stem about every 15 metres. The
retained material often contains groundwater and so the gap (expansion
joint) must be bridged to prevent water coming through to the exposed
face of the wall.

Details taken from a manufacturer's trade literature (Fig. 9.4) show
suitable waterproofing arrangements for the expansion joint. The
water-bar is also shown in cross-section and dimensioned (b). It is
designed to accommodate expansion and contraction in the walls (c); the
joint filler is soft and easily compresses when the wall expands and the
sealing compound remains plastic. The details shown may be

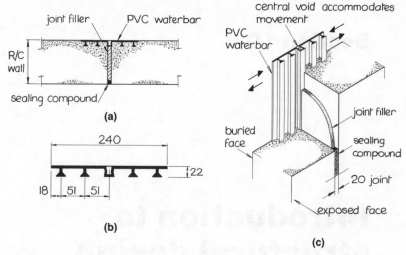

Fig. 9.4

reproduced on the working drawings and called up with the manufacturer's name and product references.

If the design team decide to change the recommended detail, it is normal to discuss this with the manufacturer. When several companies produce a similar range of products, the structural details may be prepared based on one particular system but with the note 'or similar approved' added to allow the contractor to use an alternative.

9.4 Libraries

Although many design offices keep extensive libraries of trade literature, most projects can be detailed using only a few publications, and designers and detailers may acquire a personal collection that they use regularly. It is, however, always important to check that the publication being used is the most recent.

Section 10

Introduction to structural design

10.1 Introduction

Much of the information required to detail structures is taken from the structural calculations.

The design of a building is initiated by an architect, who prepares outline schemes for the project. From this information the structural design team prepare their own schemes, identifying those elements which will be used to form the structural framework. When the basic layout is finalised, calculations are prepared which include the design of the principal elements such as floor slabs, beams, columns and foundations, as well as the design of details such as bracings and cladding support fixings. Calculations prepared by an engineer should give all the basic information needed to detail the structure.

This section describes basic principles of design used in building and civil engineering structures.

10.2 Building frames

A typical framed building structure (Fig. 10.1(a)) consists of floor slabs spanning onto supporting beams, and the beams, in turn, span between columns. In a multi-storey building loads accumulated at each floor are carried progressively down the columns into the foundations. The structure may be a skeleton of steel beams and columns supporting cast-*in-situ* or pre-cast concrete floors. Alternatively, it may be a fully

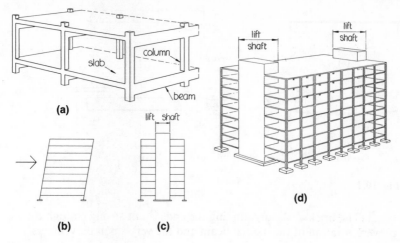

Fig. 10.1

cast-*in-situ* concrete framework of columns, beams, walls and floors. The completed structure must be capable of resisting (horizontal) wind forces which cause a racking effect (b). This may be taken by the frame, although it is more common to carry these effects in the stiff vertical elements within the frame such as lift shafts and gable walls (c, d). The completed building will have external claddings, usually of pre-cast concrete panels or brickwork, and internal partition walls of blockwork or lightweight (stud) construction. If these elements do not contribute to the strength or stability of the building, then they are referred to as non-loadbearing.

In other types of building, the external claddings and internal walls may be used to support floors and roofs and to brace the structure, giving horizontal stability. These walls are referred to as loadbearing. Structures such as industrial buildings, tanks and bridges can also be broken down into their component parts for the purposes of design.

10.3 Beams and slabs

Beams and slabs are the basic horizontal members in a structure and the simply supported beam is a commonly occurring structural element. In calculations, it is shown as a horizontal line representing the beam and vertical arrows representing the supports (Fig. 10.2(a)). When loaded, the beam 'sags', deflecting freely and rotating on the supports (b) which are described as 'pinned' or 'free'. The concrete beam standing on brick walls (c) is an example of a simply supported beam. The walls give the necessary vertical support but the bed joint cannot prevent rotation. The lintel supported on brickwork (d) is also designed as a simply supported

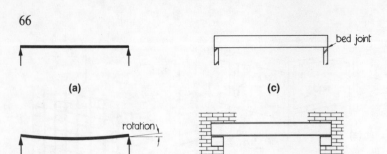

Fig. 10.2

beam. The brickwork surrounding the ends is not strong enough to prevent rotation of the loaded beam and the very small movements which take place may cause minute cracks.

The cantilever has only one support (Fig. 10.3(a)) which not only gives vertical support like the pinned joint, but also prevents rotation (b). The beam arches or 'hogs' and the support is described as 'rigid' or 'fixed'. The balcony (c) is a common example of a cantilever structure.

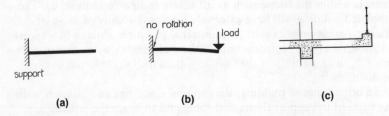

Fig. 10.3

10.4 Loadings

The complex loading arrangements that occur on a structure are defined in simplified terms to allow for practical analysis. The most common loads are described as 'concentrated' or 'point' loads and are shown in calculations as an arrow (Fig. 10.4(a)), and 'uniformly distributed' loads

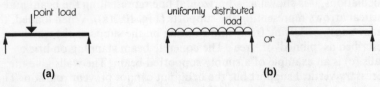

Fig. 10.4

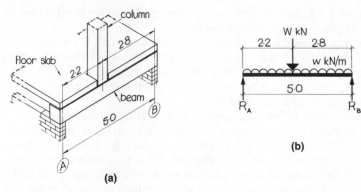

Fig. 10.5

which are uniform along the length of the beam and are shown either as a line of humps or as a straight line (b). The loads carried by each support are referred to as the 'support reactions'. They are calculated at an early stage in the design of beam and slab structures. When a beam carries several different loads, their total effect is calculated as the sum of the separate loads. In the example shown in Fig. 10.5(a) the concrete beam is supported on brick walls and carries a concrete floor slab over its whole length and a column slightly off-centre. In the calculations this is shown as a simply supported beam (b). The span is the distance between the centres of the support, taken in this case to be the centres of the brick walls. The column is a concentrated load W (kN) and the distributed load w (kN/m) is the self-weight of the beam, the slab and other effects. The loads carried by the walls are the support reactions R_A and R_B.

Loads are carried into a beam as bending moments and shearing forces. Bending moment is a flexural effect and is measured as the product of force × distance. At the point x–x on a beam (Fig. 10.6(a)) the force P acting vertically downwards at a distance l_1, causes a bending

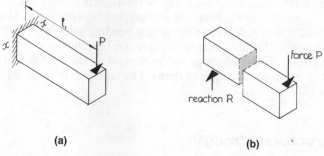

Fig. 10.6

moment of $P \times l_1$. In structural calculations, the force (or load) is usually measured in kN and the distance in metres.

Bending moment is measured in the compound unit kN m, and the value varies along a beam. Shear force is a fracturing effect which occurs across the plane of an element (b). Like bending moments, the value of shear force varies along a beam and at any point is dependent on the size and location of the applied forces.

The variation of bending moment and shearing forces is shown in simple graphical form as bending moment and shear force diagrams (BMDs, SFDs). Figure 10.7 shows details of loadings, deflected shapes, bending moment and shear force diagrams for simply supported beams and cantilevers. The maximum (design) values of bending moment and shear are given for each beam. These are more fully explained in books on structural analysis.

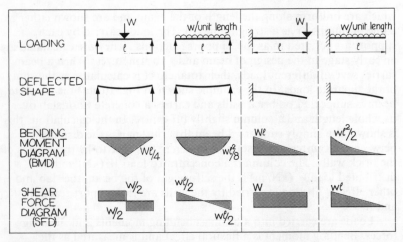

Fig. 10.7

Beams and slabs which span across several supports are described as 'continuous'. Figure 10.8 shows a beam continuous over four spans and carrying a uniformly distributed load. The beam deflects to the form shown in (b), hogging over the supports and sagging between them. The effect of continuity is to redistribute the simple bending moments ($wl^2/8$ etc.), reducing the maximum span value but introducing a support moment (c). The final design values depend on the distribution of load and the number and length of spans. The variation in shear force is also shown (d).

10.5 Structural strength

The capacity of a beam or slab to carry bending moments is referred to

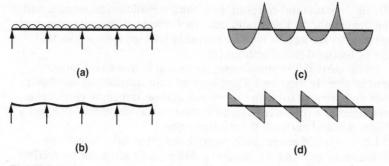

(a)

(b)

(c)

(d)

Fig. 10.8

as the moment of resistance (M_R). It is the product of the cross-sectional shape and size of the beam and the flexural strength of the material.

The simplest beams and slabs are normally designed for the maximum bending moments and shear effects only. For more complex flexural elements the design may be matched to the variations in bending moment and shear force. Steel sections may have additional plates welded to the flanges to strengthen the section to carry local high moments, and, in reinforced-concrete structures, it is normal practice to vary the amount of reinforcement used in both flexure and shear, in accordance with the bending moment and shear force diagrams. Preparing drawings which show the layout of steel reinforcement in concrete structures is an important part of structural detailing.

10.6 Columns and walls

Columns and walls form the vertical load-carrying elements in structures.

Concrete columns are normally cast as square or rectangular sections (Fig. 10.9(a)) and concrete walls (b) are slender plate-like

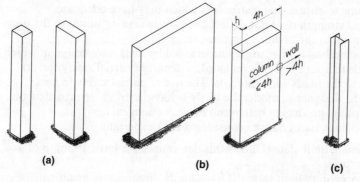

(a)

(b)

(c)

Fig. 10.9

elements. In structural calculations a concrete column becomes a wall when the length/thickness ratio exceeds 4. Steel columns are commonly referred to as stanchions and are normally selected from a standard range of I-shaped rolled sections (c).

Loads carried by columns may be caused by the beam support reactions already described. Load applied concentrically to a column causes compressive stress in the section. A steel stanchion carries the stress directly; a reinforced-concrete column may carry the load partly in the concrete and partly in the reinforcement.

Loads applied eccentrically to columns cause axial forces and bending moments, and the structural design must allow for both effects.

10.7 Structural calculations

At an early stage in the design of a structure, calculations are required to determine the materials to be used in the construction and the sizes of the elements which must be adequate to carry the anticipated loads. In theory, it should be possible to load structural materials right up to their yield strengths. In practice this would be very unwise because a small overload could result in structural failure. The problem is overcome by applying suitable factors of safety to ensure that the effects of applied loading remain below the yield strength by an adequate margin. Some structural design is carried out using a *permissible* (or working) *stress* method in which the effects of the applied loading are matched directly with a permissible stress (strength) for the material. In general, permissible strength may be expressed as

$$\frac{\text{yield strength}}{\text{factor of safety}}$$

Different factors of safety are used for different materials and 'overstress' is often allowed for short-duration effects such as wind-loading. Where appropriate, calculations are also made to check deflections to ensure that a structure which may have adequate structural strength does not deflect excessively causing alarm to the users and damage to the building finishes, etc.

This method is now largely superseded by *limit state* design in which the single factor of safety is replaced by separate partial safety factors applied to the loads and materials. The use of partial safety factors allows the designer to model the likely behaviour of the loaded structure more closely, producing better and more economical designs.

Two limit states may be considered for the design:

(i) ultimate limit state (ULS)—this determines the basic strength of a structure.
(ii) serviceability limit state (SLS)—checks the structure against excessive deflections, cracking and other durability effects.

Reinforced concrete, structural steelwork and masonry may be designed using limit state design methods. Structural timber is still designed using permissible stresses.

10.8 Calculation sheets

Hand and computer calculations may be used in the design of a structure. Hand calculations are normally prepared on pre-printed A4 sheets (Fig. 10.10). The basic analysis and design are shown in the central section and the wide right-hand margin is used to summarise the results of the calculations. For example, in the design of a reinforced concrete floor slab, the margin might be used to list details including the slab thicknesses and the reinforcement required (b). This information is used in detailing the working drawings.

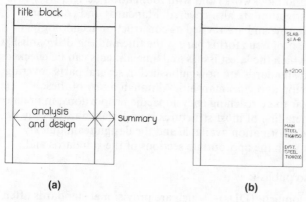

(a)   (b)

Fig. 10.10

Computers are also used in structural design, usually for repetitive calculations or to solve a major design problem such as the analysis of a complex framework, or a bridge deck, and often the design team use computer design data in conjunction with hand calculations.

10.9 Technical design standards

Building and civil engineering works are designed in accordance with the requirements and recommendations of technical standards appropriate to the particular type of structure. Referring to these documents and applying the information they contain is an important part of the design and detailing of structures.

Some standards are in constant use and the design team may keep their own copies for immediate reference; others are required only occasionally and may be kept in an office library.

10.10   British Standards Institution (BSI)

The British Standards Institution is the principal source of design and technical standards used in the construction industry. They publish:

British Standard Specifications—normally referred to as British Standards or BSs.
British Standard Codes of Practice—referred to as Codes of Practice or CPs

There are British Standards and Codes of Practice dealing with the design of structures in reinforced concrete, steelwork, timber and masonry (brick and blockwork) and with foundation design. Activities such as site investigation are also covered. Standards and codes are under constant review and are revised as construction technology advances. The year of issue forms part of the title and the BSI publish a yearbook from which the latest issues and amendments can be checked.

Some design standards are now published in several parts covering design, construction and workmanship. Although many of these documents appear to be reaching encyclopaedic proportions, in practice, the design and detailing of most structures is undertaken using only a small part of the information available and the design team quickly become familiar with the appropriate sections of the standards and codes.

The BSI also publish:

Drafts for Development (DDs)—which are provisional standards often produced when guidance is urgently needed. They may subsequently be converted into British Standards or withdrawn.
Published Documents (PDs)—which are advisory and informative. They do not have the same status as Standards.

10.11   Department of Transport (DTp)

The Department of Transport publish a range of design documents specifically for highway bridges—BAs (bridge advice), BDs (bridge design) and BEs (bridge engineering). They are identified by a number and the year of publication. For example, BE1/78, *Design Criteria for Footbridges and Sign/Signal Gantries*. There are documents covering all aspects of bridge design and detailing and some are read in conjunction with the appropriate British Standards. An annual index is published.

10.12 Building Regulations

Building work in Great Britain must be carried out in accordance with the Building Regulations. These are rules for good building practice which are enforceable by the local authorities through building control offices.

Practical guidance in the interpretation of the Regulations is given in a set of 'approved documents' which include a section on structures. The documents refer to a number of British Standards and Codes of Practice. The use of these design codes gives automatic compliance with appropriate aspects of the Building Regulations.

10.13 Building Research Establishment (BRE)

The Building Research Establishment undertakes research into many aspects of building construction including materials, construction methods and safety problems. They publish many technical papers including *BRE Digests*. These are brief technical notes which are regularly used in design offices.

10.14 Agrément certificates

The British Board of Agrément (BBA) is an independent government body which tests and assesses proprietary products used in the construction industry. The BBA issue Agrément certificates for tested products which meet the required standards (Building Regulations, etc.) These certificates may be supplied with a manufacturer's catalogue as an indication of the reliability and suitability of the product.

10.15 Technical advisory organisations

There are numerous technical advisory organisations serving the construction industry, including:

Concrete—Concrete Advisory Service (Concrete Society)
Steel—The Steel Construction Institute (SCI); formerly CONSTRADO
    The British Constructional Steelwork Association Ltd (BCSA)
Timber—Timber Research and Development Association (TRADA)
Brickwork—Brick Development Association (BDA)

All publish literature and offer technical information about design and detailing problems and the use of materials.

Part 2

Section 11

Reinforced concrete

11.1 Introduction

Reinforced concrete is used extensively in building and civil engineering
construction. Concrete can be cast into a variety of shapes to create
structures which have great strength and durability. Building frames are
usually masked by brickwork or similar claddings, but for other
structures, the concrete may be left exposed and special casting
techniques and surface treatments used to give a distinctive finish.

Figure 11.1 shows: (a) a reinforced-concrete building frame nearing
completion, and (b) a concrete bridge at the deck construction stage.

The finished building is brick clad; bricklaying is already under way
at ground level, but on the bridge the concrete is fully exposed and
particular care is taken to achieve a good surface finish.

These structures are cast *in situ* and need extensive temporary
support works during construction. Alternatively, individual elements
such as beams and columns may be pre-cast in a workshop, brought to
site and assembled with a minimum of temporary works. With any form
of construction which involves casting, it is not possible to check the
quality and appearance of the work until it has hardened and the mould
is removed. Deficiencies in surface finish can sometimes be repaired, but
understrength concrete is rarely acceptable and must usually be
demolished and re-cast. Such corrective action is very expensive.

The design and detailing of concrete structures must exploit the
advantages as well as catering for the limitations of the materials used in
construction. Accepted standards for quality and workmanship are

Fig. 11.1 (a)

embodied in specifications and interpreted on site by custom and practice.

This section describes the principal characteristics of concrete as they relate to structural detailing.

11.2 Design codes

The principal codes used in the design of reinforced concrete works are:

BS8110: 1985: Structural use of Concrete—this is a limit state code enabling structures (principally buildings) to be designed for ultimate and serviceability conditions.

BS5400: Part 4: 1984: Steel, Concrete and Composite Bridges. Code of Practice for the design of concrete bridges—this is also a limit state code; it was prepared specifically for the design of bridges.

BS8007: 1987: Design of concrete structures for retaining aqueous

Fig. 11.1 (b)

liquids. This is a limit state code which is used mainly in the design of water-retaining structures such as tanks.

11.3 Construction sequence

The sequence of operations required to construct a simple reinforced-concrete structure starts with the construction of *formwork* (shuttering). This is the mould in which the concrete will be cast. Basic formwork is usually a timber construction of 20 mm thick plywood sheets nailed to a framework of 100×50 mm softwood and supported on adjustable tubular steel props or frames. The plywood surface is either oiled, with a special mould oil, or otherwise prepared, to prevent the hardened concrete from sticking to it. Steel fixers place and secure reinforcing bars in the shutter in positions required by the structural design and then the concreting gang pour wet concrete in layers distributing it with shovels so that it surrounds the steel and fills the shutter. As each layer is placed, a poker-vibrator, usually 40–50 mm diameter, is inserted into the concrete. This finally distributes and settles the concrete, shaking the trapped air to the surface. The process is known as *compaction*. The exposed surface of the concrete is levelled with a tamping board and is then left or further smoothed with a wooden float or steel trowel. A temporary sealing membrane, usually polythene sheet or chemical spray, is applied to the exposed concrete surface so that the initial setting and hardening (curing) can take place without excessive moisture loss due to evaporation. Without this protection there could be severe surface cracking due to shrinkage and a loss of concrete strength. As the concrete hardens sufficiently, the formwork is removed (stripped) progressively, starting with the side and then the soffit shutters. The props are normally kept in place until the concrete structure is self supporting. The same shutters may be re-used several times before the plywood lining requires replacement.

11.4 Construction drawings

This construction sequence is carried out by a contractor working from drawings prepared by the design team. For each element in the structure, such as a wall or floor slab, two sets of information are required and these are usually shown separately as a layout or general arrangement drawing and reinforcement details.

A summary of the information required on these drawings is shown in Fig. 11.2.

For very simple elements, the information is sometimes combined on a single drawing.

The quality of the materials to be used and the control of construction procedures are detailed in the structural specification which must be compatible with the drawings.

11.5 Concrete

In a normal structure, different concretes may be used for different parts of the work, but they are all basically a blend of cement, coarse and fine aggregates and water.

Ordinary Portland cement (OPC) is the most commonly used of the cements (about 90% of UK supply). The characteristic grey colour of OPC derives from iron compounds in some of the ingredients and the name Portland is taken from the similarity of the hardened cement to limestone rock at Portland in Dorset. White Portland Cement is used when an exposed white finish is required. It is made in specially controlled conditions using materials which are substantially free from the grey pigmenting compounds.

Rapid-hardening Portland cement (RHPC) is similar to OPC except

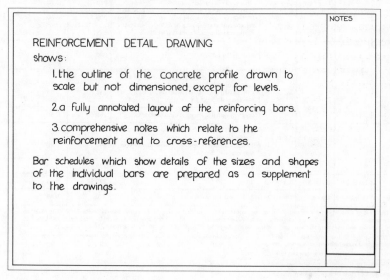

(a) layout or general arrangement drawing

(b) reinforcement detail drawing

Fig. 11.2

that it is more finely ground. Although it sets at a similar rate, it acquires strength more quickly, making it useful in situations where the concrete requires early strength to allow work to progress. The final strength is similar to that of OPC concrete.

Sulphate-resisting Portland cement (SRPC) is used mainly in below-ground works where sulphates are present in the soil or groundwater. Information on sulphate content and pH (acidity) values should be included in the site investigation report for a project. The sulphate content is expressed as a percentage of the soil sample or alternatively in grammes per litre in ground water, or a 2:1 water:soil extract. Table 17 of BS8004: 1986: Code of Practice for Foundations, lists the type and quantity of cement to be used in concrete for different sulphate concentrations. The table is shown in Fig. 11.3.

Table 17. Concrete exposed to sulphate attack

NOTE. The recommendations are for concrete in a near-neutral groundwater; for acid conditions see Gutt and Harrison (1977).

Class	Concentration of sulphates expressed as SO_3			Type of cement		Dense fully compacted concrete made with aggregates complying with BS 882 or BS 1047	
	In soil		In ground-water			Minimum cement content*	Maximum free water/cement* ratio
	Total SO_3	SO_3 in 2:1 water:soil extract					
	%	g/L	g/L			kg/m^3	
1	Less than 0.2	Less than 1.0	Less than 0.3	Ordinary Portland cement (OPC) Plain Rapid hardening Portland cement concrete† (RHPC), or combinations of either Reinforced cement with slag‡ or p.f.a.§ concrete Portland blastfurnace cement (PBFC)		250 / 300	0.70 / 0.60
2	0.2 to 0.5	1.0 to 1.9	0.3 to 1.2	OPC or RHPC or combinations of either cement with slag or p.f.a. PBFC		330	0.50
				OPC or RHPC combined with minimum 70 % or maximum 90 % slag‖ OPC or RHPC combined with minimum 25 % or maximum 40 % p.f.a.¶		310	0.55
				Sulphate resisting Portland cement (SRPC)		290	0.55
3	0.5 to 1.0	1.9 to 3.1	1.2 to 2.5	OPC or RHPC combined with minimum 70 % or maximum 90 % slag OPC or RHPC combined with minimum 25 % or maximum 40 % p.f.a.		380	0.45
				SRPC		330	0.50
4	1.0 to 2.0	3.1 to 5.6	2.5 to 5.0	SRPC		370	0.45
5	Over 2	Over 5.6	Over 5.0	SRPC plus protective coating**		370	0.45

*Inclusive of content of p.f.a. or slag. These cement contents relate to 20 mm nominal maximum size aggregate. In order to maintain the cement content of the mortar fraction at similar values, the minimum cement contents given should be increased by 50 kg/m^3 for 10 mm nominal maximum size aggregate and may be decreased by 40 kg/m^3 for 40 mm nominal maximum size aggregate.

†When using strip foundations and trench fill for low rise buildings in class 1 sulphate conditions, further relaxation in the cement content and water/cement ratio is permissible.

‡Ground granulated blastfurnace slag (see BS 6699).

§Selected or classified pulverized-fuel ash complying with BS 3892.

‖Per cent by mass of slag/cement mixture.

¶Per cent by mass of p.f.a./cement mixture.

**See CP 102.

NOTE 1. Different aggregates require different water contents to produce concrete of the same workability and therefore a range of free water/cement ratios is applicable to each cement content. In order to achieve satisfactory workability at the specified maximum free-water/cement ratio it may be necessary to increase the cement content above the minimum specified.

NOTE 2. Within the limits specified in this table, the use of p.f.a or slag in combination with sulphate resisting Portland cement (SRPC) will not give lower sulphate resistance than combination with cements complying with BS 12.

NOTE 3. If much of the sulphate is present as low solubility calcium sulphate, analysis on the basis of a 2:1 water extract may permit a lower site classification than that obtained from the extraction of total SO_3. Reference should be made to BRE Current Paper 2/79 for methods of analysis and to BRE Digests 250, 275 and 276 for interpretation in relation to natural soils, fill and hardcore.

Fig. 11.3

This information is also given in other documents, including BS8110 and *Building Research Station Digest* No.250.

11.6 Aggregates

The aggregates used in structural concrete are usually naturally occurring materials such as rocks and gravels, although artificial aggregates manufactured from materials such as pulverised fuel ash (pfa) are also used in certain applications.

Dense strong concrete is made from a blend of aggregates with different sized particles, which bind together to form a compact matrix. Natural aggregates are defined in BS882: Coarse and Fine Aggregates from Natural Sources. The Standard defines a range of aggregate sizes between 75 mm and 150 microns (0.15 mm) measured as the hole size of a square mesh through which the materials can pass. Aggregate blends are graded by controlling the amount of each particle size present. Coarse aggregates are referred to by the maximum particle size, for example 20 mm, 10 mm, and they have a minimum particle size of 5 mm. Fine aggregates contain particles which are less than 5 mm. They are sub-divided into four zones, 1–4, by varying the proportions of fine and very fine materials present. This allows additional control in the production of concrete mixes of different characteristics. 'All-in' aggregate defined in the Standard is a combination of coarse and fine materials. It is suitable for low-grade concrete.

11.7 Water

The water used for making concrete must be free from impurities. Mains water is normally suitable; sea water can be used under certain circumstances but not for concrete which is to be reinforced because it contains salts which cause corrosion.

11.8 Concrete strength

Cement, aggregates and water can be mixed to produce concretes of many different types and strengths. Strength is measured as the stress (in N/mm^2) required to crush sample cubes of the hardened concrete. This value defines the *grade* or *class* of the material. Concretes for structural works are normally made in a range of strengths. Specific values are listed in design codes and standards including BS5328: 1981: Methods for Specifying Concrete, including ready-mixed concrete. The concrete may be reinforced or unreinforced. Unreinforced, it is referred to as *mass* or *plain* concrete. Commonly used concretes are listed in Fig. 11.4. Concretes of grade C7.5, C10 and C15 are made with an 'all-in'

Grades of concrete	
Unreinforced	**Reinforced**
C7.5	C25
C10	C30
C15	C35
C20	C40

Fig. 11.4

aggregate. Concretes of grade C20 and stronger are made with a blend of coarse and fine aggregates. C25 is the lowest grade that should be used for reinforced concrete made with normal-weight aggregates such as crushed rocks and gravels.

In addition, designers often specify class E concrete as a cheap low-strength mix used mainly as a bulk-fill material for making good in foundation works, protecting drain runs and so on. The mix is 1:8 or 1:10 (cement:all-in aggregate) measured by volume. The cube strength is unspecified but is unlikely to be more than 15 N/mm². Class E concrete should never be used in exposed situations or for reinforced-concrete work.

11.9 Water/cement ratio

The setting and hardening of wet cement is a chemical reaction, and newly made concrete must contain sufficient water to enable this to take place and also to produce a free-flowing mix which can be placed and compacted.

The amount of water used in the mix is related to the cement content as the water/cement (w/c) ratio:

$$\text{Water/cement ratio} = \frac{\text{weight of water}}{\text{weight of cement}}$$

The ratio is critical and should be a minimum consistent with the production of properly compacted concrete. Increasing the ratio (and therefore the amount of water present) improves the free-flow characteristics of the mix, but the strength of the hardened concrete may be substantially reduced, and the shrinkage, which always occurs when concrete dries out, may become excessive. On the other hand, if the water/cement ratio is too low, proper compaction may not be possible. This may also reduce the strength of the hardened concrete. Typical values of water/cement ratio for structural concretes are 0.5–0.6.

The flow characteristics of the wet concrete are referred to as the *workability* and are dictated by the size of the concrete profile and the ease with which the concrete can be poured, the amount of reinforcement and the method of compaction. Low-workability concretes are appropriate for easily accessible, large, lightly reinforced elements. Small, heavily reinforced sections may require a freer-flowing, high-workability concrete. An indication of the workability of a concrete may be obtained from a standard hand test.

11.10 Admixtures

The characteristics of a concrete may be modified by the use of admixtures. These are materials added to a mix in controlled quantities to produce a particular effect, for example:

Accelerators—speed up the setting of concrete.
Retarders—slow down the setting of concrete.
Plasticisers—improve the concrete's workability.
Air-entrainers—add and then control the dispersal of tiny bubbles
of air in the mix, improving workability and
producing a more durable concrete.

Admixtures may be specified by the design team but are more commonly used by contractors as an aid to construction. Careful control is required in their use.

11.11 Concrete supply

The volume of concrete required in a structure is rarely sufficient to justify the expense and complication of setting up a mixing plant on site. Instead it is supplied by a specialist ready-mix contractor who makes the concrete at a permanent batching plant in the area and then delivers it to the site in a rotating-drum lorry. About 75 per cent of the concrete used in construction is supplied in this way. On-site mixing plants are used when large quantities of concrete are required for structures such as dams, or when ready-mixed concrete is not easily obtainable.

Concretes can be made with specific strengths and workabilities. The process of designing a blend of materials which will achieve a required performance is referred to as *mix design* and it may be undertaken either by the design team or the concrete contractor. Normally, the design team supply a performance specification and leave the detailed design to the contractor. In BS8110, this is referred to as a *designed* mix. Alternatively they may retain control of the design and specify a *prescribed* mix. For highway structures the Specification for Highway Works refers to *designed* mixes.

Details of the concrete mixes must be given on the structural drawings and in the specification. On the drawings, details of the concrete mix are usually restricted to simple notes, typically:

Concrete shall be grade C30 using ordinary Portland cement and 20 mm maximum aggregate.

Full information about the materials, production, quality control and transportation of the concrete are then given in the structural specification. The concrete supplier designs a suitable mix to these requirements and prepares a certificate which is submitted to the design team for approval before concrete is supplied to the site.

If a prescribed mix is to be used, the letter P is added to the description, for example C35P.

11.12 Example

The selection of suitable concretes is described for a two-storey cast *in situ* concrete-framed building (Fig. 11.5). The foundation soil is a dense gravel and no groundwater is present. Tests on soil samples show that the sulphate content is less than 0.1 per cent. The building stands on individual pad foundations. A large-diameter pipe is laid in a trench adjacent to the foundation. In the completed building, the ground floor columns will be exposed and it is an architectural requirement that they should be cast with white Portland cement. The structure would be designed to BS8110.

Referring the soil sample analysis to table 17 of BS8004 (or

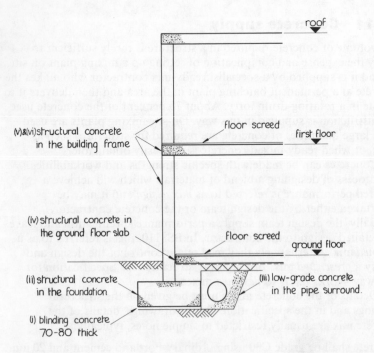

Fig. 11.5

table 6.1 of BS8110) shows that the sulphate content is very low and all concrete in contact with the soil may be class I using OPC.

(i)   The pad foundation is reinforced and a clean and rigid working surface is required on which to fix the cages of reinforcement and construct the shutters. This surface is made by pouring a layer of *blinding concrete* usually 70–80 mm thick across the whole area of foundation. A grade C7.5 concrete with 20 mm maximum aggregate would be suitable.

(ii)   Pad foundations are normally simple square or rectangular blocks of concrete. Typical sizes are from $1.5 \times 1.5 \times 0.5$ mm thick to $4 \times 4 \times 1$ m thick. Access to place and compact the concrete is usually good and the foundation would probably be lightly or moderately reinforced. If the excavation will stand vertical, the concrete may be cast directly against the soil, saving the cost of a foundation shutter. A grade C35 concrete with 20 mm maximum aggregate and low or medium workability would be suitable.

(iii)   The pipe is large and relatively near to ground level and the layer of hardcore which is placed over it for the ground floor slab must be compacted, probably with a vibrating roller. To prevent the pipe from being damaged during this operation, it will be surrounded with grade C20 concrete giving an adequate economic protection.

(iv)   The ground floor slab is cast on a compacted hardcore base and reinforced with sheets of fabric. The fabric is quickly and easily fixed and it is not therefore necessary to cast a blinding concrete working surface. Instead, the compacted hardcore is blinded with a layer of fine stone dust or sand which fills the voids and can then be rolled to form a smooth, flat surface. A plastic sheet membrane (typically 1000 gauge polythene) is laid on the blinding. The reinforcing mesh may be fixed at this stage or it can be placed as the slab is concreted. A grade C25 or C30 concrete with 20 mm maximum aggregate and low or medium workability is suitable.

(v)   The elements forming the superstructure are generally made as slender as possible. This results in moderate or heavily reinforced sections often with areas of congested reinforcement. A grade C35 or C40 concrete with 20 mm maximum aggregate and medium workability would be suitable. The need for white cement in the ground-floor columns must be stated on the drawings. The externally exposed concrete would be grade C40.

(vi)   The top of the cast *in situ* concrete floor is normally finished by tamping which produces a rough, ribbed surface. The smooth floor required in the finished building is achieved by laying a 40 or 50 mm skin of sand:cement screed (usually 3:1) which can be floated to give a surface smooth enough to take thermoplastic tiles or carpets. In this example, the screed does not contribute to the strength of the floor, and it is referred to as non-structural. Full details of the screed would be shown on the architect's drawings; but for completeness, it can also be indicated on the structural details. Screed can also be used to give added strength to the floor construction (structural screed) and then it should be fully detailed on the structural drawings and in the specification.

Basic details of the concretes to be used are given in the notes box on the drawings. On a typical project a number of layout drawings are prepared, each showing different parts of the structure. Suitable 'concrete' notes for the foundation and ground works drawing would be:

1. Blinding concrete shall be grade C7.5 with 20 mm maximum aggregate.
2. Concrete in structural foundations shall be grade C35 with 20 mm maximum aggregate.
3. Concrete for the pipe surround shall be grade C20 with 20 mm maximum aggregate.

Full details of these mixes, including the type of cement, minimum cement content and water/cement ratio, would be given in the accompanying structural specification.

For any structure, the choice of concrete strengths, workabilities and so on must be based on the particular structural requirements which have been assumed by the design team and on the conditions in which the structure is built.

11.13 Surface finish

Concrete is cast in formwork (shuttering) which is kept in place until the concrete has been hardened sufficiently to be self-supporting. The shutter can then be removed. If the wet concrete is properly compacted, the cast surface will mirror the surface of the shutter in every detail.

The quality of the shutter very largely determines the quality and surface profile of the finished work. Concrete which is to be exposed in the completed structure may require a high-quality surface finish while the appearance of concrete which will be buried in foundation works is unimportant and a lower standard is acceptable. The structural drawings must include details of the surface finishes required.

Simple formwork is normally constructed from 20 mm plywood nailed to softwood framing and supported by adjustable tubular scaffold props. There are also many other proprietary systems such as flat panels made as metal frames with replaceable plywood faces, and tunnel-shaped moulds for casting certain types of floor. Wet concrete imposes considerable forces on the shuttering which must be constructed to remain stable and rigid throughout the casting and curing period. The joints must be tight to prevent uncontrolled seepage of fine materials which would affect the durability and appearance of the finished concrete.

The different qualities of surface finish are normally described in the structural specification and given reference letters or numbers which are then used on the drawings to identify the different finishes required throughout the structure. For example, the Specification for Highway Works describes a range of surface qualities which are common to most structural works. Two types are described:

(a) Formed surfaces (F series)—which are cast against a shutter.
(b) Unformed surfaces (U series)—which are the exposed surfaces of slabs, tops of walls and so on, finished by tamping and trowelling.

The formed surfaces start with class F1. This is a minimum standard which includes requirements that the formwork is solidly built and does not allow seepage. This finish would be adequate for the sides of the pad foundation in the building (Fig. 11.5). Reclaimed plywood is often used when an F1 finish is specified. The standard F series finishes progress (F2, F3, F4, etc.) to include high-quality 'fair-faced' surfaces suitable for exposed concrete work such as the ground-floor columns in the building. Identifying formed surfaces by F numbers can be extended to include special (non-standard) finishes by using the next number in the sequence (F6, F7, etc.). The references are used on the drawings to identify the concrete surfaces and are related to descriptive clauses in the specification. They would include textured concrete finishes cast against rough-sawn timber, and proprietary profiled shutter liners.

Unformed surfaces start with class U1 which is a basic levelled surface produced by tamping. A U1 finish would be suitable for the floor slabs in the building, and the classification progresses to surfaces which are first tamped and then floated or trowelled to produce a specific finish (U2/U5). Other special unformed finishes could be specified by using the next number, U6, U7, etc., and writing the appropriate descriptive clauses in the specification.

On the drawings the different surface finishes required must be identified against each concrete face and this is often done with a circled pointer ⊙ . Alternatively, when one or two finishes are used extensively, notes may be used, for example:

1. Formed surfaces
   (a) All finally buried concrete shall have a class F1 finish.
   (b) All finally exposed concrete shall have a class F4 finish.
2. Unformed surfaces.
   All unformed surfaces shall have a class U2 finish.

A further note for a test panel could be added:

3. Sample panel(s) shall be made for the exposed, board mark finished concrete (class F6). The size and thickness of the panel(s) shall be $1.0 \times 1.0 \times 0.25$ m thick and the method of construction shall be truly representative of that which will be used on the main structure.

Surface finishes are also described in BS8110.

Section 12

Reinforcement

12.1 Introduction

A standard range of bar types and sizes is available for use in reinforced-concrete structures. This section describes the physical properties of those most commonly used and shows typical reinforcement arrangements for the basic elements of a structure. It does not include the steels used in pre-stressed concrete or any of the range of special steels such as Macalloy bar.

12.2 Types of steel

The steel commonly used in reinforced concrete structures is of two types:

(a) hot-rolled—which may be
    (i)   mild steel, or
    (ii) high-yield steel;
(b) cold-worked—high-yield steel.

12.3 Range of reinforcement

Bars are made in a range of diameters or nominal sizes, usually from 8 to 40 mm. Bars of size 6 and 50 mm are also available. Nominal sizes are based on the diameters of plain round bars. Certain types of bar have more complicated cross-sections but they are rolled to the same range of areas as the round bars and are identified by the same nominal sizes.

Reinforcement manufacturers normally supply bars in 12 m lengths, although longer bars (up to 18 m) can be supplied by special order. Reinforcement is also available in the form of pre-fabricated sheets made by welding small-diameter bars together to form a grid. The sheets are called *fabric* and the spacing of the longitudinal and cross-wires is called the *mesh*. Fabrics are available in a range of bar and mesh sizes. The standard size for the sheets is 4.8×2.4 m and the fabric can also be produced in 2.4 m wide rolls.

12.4 Specification

The specification for the steels is covered by British Standards which include requirements for:

(a) chemical composition
(b) tensile strength
(c) ductility
(d) bond strength
(e) weldability
(f) cross-sectional area

12.5 Chemical composition

The steels are alloys of iron, carbon and a number of other elements such as magnesium, nickel and chromium. The precise composition varies for the different types but is normally about 98 per cent iron and 2 per cent carbon and the other elements.

Hot-rolled reinforcement is made by rolling steel of different strengths to produce profiled reinforcing bars. Cold-worked reinforcement is made in the same way, but the profiled bars are subsequently cold-twisted to form spirals, which stretches the crystalline structure of the steel and increases the tensile strength.

12.6 Tensile strength

The tensile strength of steel may be determined by testing a standard specimen by applying a tensile load in increasing increments. As each increment of load is applied, the specimen stretches and the resulting extensions are measured. The test is normally continued until the specimen fails. Graphs plotted of load/extension are used to determine the tensile strength and other characteristics.

A typical load/extension graph for hot-rolled steel is shown in Fig. 12.1(a). From O to A the load/extension graph is straight which means that the extension of the specimen is proportional to the applied load. If the load is doubled, the extension doubles. O–A is called the *elastic range* of the metal and if the load were removed, the specimen would revert to its original length. In a working structure the stresses in the reinforcement must be within the elastic range. The point A indicates the end of the elastic range. It is called the *yield point* and the stress in the metal is the *yield stress*. The permitted design strength for hot-rolled reinforcing steel is derived from this stress.

Beyond A, the test specimen lengthens without any increase in applied load. Strain hardening takes place in the metal and the specimen stabilises at B. Further increases in the applied load (B–C–D) cause further extension and strain hardening. Eventually a maximum load is reached at which the compensating effect of the strain hardening can no longer control the extension and the specimen 'necks' and breaks (E). The point D indicates the maximum tensile strength of the metal and point E gives the *elongation at break*.

A typical load/extension graph for cold-worked steel is shown in Fig. 12.1(b). The initial cold-working process stresses the steel beyond its hot-rolled yield point (A) and causes permanent distortion. This effectively eliminates section A–B–C of the hot-rolled graph and the elastic range merges with the C–D–E part of the curve. Under test, cold-worked steels have no obvious yield point. Instead an equivalent point is defined as the 0.2 per cent proof stress, which is the stress which produces a permanent extension in the steel of 0.2 per cent.

The British Standards specify 'characteristic strengths' for the different types of reinforcement. These are the strengths used in limit state design and they are effectively equal to the yield strengths. A small

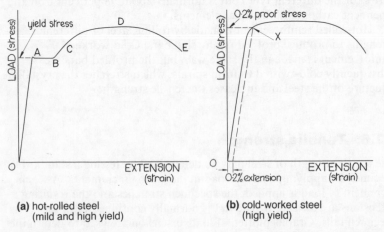

(a) hot-rolled steel
(mild and high yield)

(b) cold-worked steel
(high yield)

Fig. 12.1

tolerance is allowed for. Most commercially available reinforcement has yield strengths well above the specified values.

12.7 Ductility

Reinforcing steels must be ductile and not brittle. If a structure is accidentally overloaded, the reinforcement must be capable of considerable permanent stretching before finally breaking, because this would give early warning of failure. It would be dangerous to use steels which are brittle because, in an overloaded condition, they would be liable to sudden, unpredictable failure.

Bars are bent to specific radii for assembly into the cages and mats used in reinforced-concrete structures. The bending severely cold works the metal, and there are two ductility tests. The bend test checks the capacity of bars to withstand bending to a very small radius, and the rebend test checks the ductility of the bars after they have been cold-worked by bending.

12.8 Bond strength

To work properly, reinforced concrete relies on the grip, or bond, between the reinforcement and the concrete. Bond is measured as a stress in N/mm^2 and its strength depends on the surface profile of the bars and the strength of the concrete.

When reinforcement is supplied to site, it must be completely cleaned of millscale which is a waste product of manufacture. It may be supplied with slight surface rusting, but it must not be severely rusted, nor must the surface be contaminated with chemicals or grease. Only clean or lightly rusted steel can form the necessary bond with the concrete.

12.9 Weldability

The British Standards state that reinforcement may be welded provided that the carbon equivalent value (CEV) of the metal does not exceed 0.51 per cent. This percentage is made up of the carbon content plus fractions of the other minor elements such as magnesium, nickel and chromium. All UK-produced reinforcement fulfils this requirement although the welding methods used must be carefully controlled.

12.10 Bar types

Hot-rolled mild steel bar is manufactured to the requirements of BS4449. The specified characteristic strength of the steel (f_y) is 250 N/mm² and it is made from round bar to the preferred diameters:

8, 10, 12, 16, 20, 25, 32 and 40 mm.

Bars of 6 and 50 mm diameter are also obtainable. Although the surface of the bars is smooth, the necessary steel/concrete bond will take place when the concrete shrinks onto the steel. Bond strengths for mild steel bars cast in different grades of concrete are tabulated in design codes.

High-yield bar is manufactured to the requirements of:

(a) BS4449 for hot-rolled steel,
(b) BS4461 for cold-worked steel.

The specified characteristic strength of the steel (f_y) is 460 N/mm², which is about 1.8 times the value for mild steel. It is normally only slightly more expensive than mild steel and measured as a cost per unit strength, is clearly more economic for situations where the reinforcement is highly stressed. There are cases where high strength is not required and mild steel is more appropriate. Ribbed high-yield reinforcement is made to the same sizes as the mild steel bars.

Figure 12.2(a) shows a range of hot-rolled, high-yield bars. The surface of the bars is heavily ribbed to enable a greater concrete/steel bond to occur than is possible with the smooth mild steel reinforcement. The bars have been hot rolled to the form shown and the longitudinal rib is straight. Small identification markers are also rolled onto these bars but are not visible on the photograph.

Figure 12.2(b) shows a range of ribbed cold-worked, high-yield bars. The cold-working is easily identified by the longitudinal rib which is twisted to a spiral.

Bond strengths for high-yield steel are defined by type:

Type 1—when the current bond strengths were introduced, manufacturers were producing square twisted reinforcement, made by cold-twisting smooth surfaced square-section bar to form a spiral. This steel had a type 1 bond classification and the relevant bond strengths are tabulated in the design codes.

Type 2—this is appropriate to all high-yield bars currently manufactured in the UK. The deformed surfaces give bond characteristics which are 20 to 30 per cent better than for a type 1 bar. The type 1 bar fell into disuse and is no longer manufactured in the UK.

Mild steel bar is circular in cross-section. The shape of high-yield bar is more complicated. The core is circular but the peripheral ribs contribute to the total cross-sectional area. One range of standard areas

Fig. 12.2 (a)

(b)

based on circular mild steel is used for all types of reinforcement. For example, the cross-sectional area of a 20 mm circular bar is 314 mm²

$$\frac{\pi}{4} D^2 = \frac{\pi}{4}(20)^2 = 314 \text{ mm}^2$$

Figure 12.3 shows 20 mm mild steel and high-yield bars. The two bars have the same area but different shapes. Mild steel bars are normally referred to by diameter in millimetres; high-yield bars are identified by their nominal size, which is the diameter of a circular bar of the same cross-sectional area.

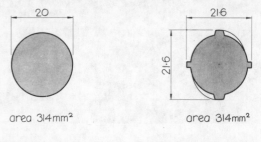

(a) 20 mm mild steel bar   (b) 20 mm high-yield bar

Fig. 12.3

On the reinforcement drawings mild steel bars are identified by the letter R followed by the diameter, thus R16 = 16 mm mild steel bar. Type 2 high-yield bar is identified by the letter T, thus T20 = 20 mm high-tensile bar. The letter X is used to identify bars not covered by R or T. The properties of type X bars must be defined in the design specifications.

12.11 Fabric

Fabric reinforcement is manufactured to the requirements of BS4483. It can be made with any combination of bar diameters and spacings, but the Standard includes a table of 'preferred fabrics' and these are used for the majority of designs. There are four types each, identified by a letter:

A—*square mesh fabric:* has bars which are the same diameter and spacing in each direction. This fabric is often used in slabs and walls where only light reinforcement is required.

B—*structural fabric:* has main wires at 100 mm centres and cross-wires at 200 mm centres. It can be used in slabs where designed reinforcement is needed.

C—*long mesh fabric:* has main wires at 100 mm centres and cross-wires at 400 mm centres. It can also be used where designed reinforcement is needed.

D—*wrapping fabric:* is very light section, square mesh fabric used mainly as a wrap-around reinforcement on structural steelwork which is to be concrete cased for fire resistance.

Type A, B and C fabrics are made from hard-drawn mild steel wire which has a characteristic strength (f_y) of 485 N/mm², or from cold-worked, high-yield bars. Type D fabric is made from drawn mild steel wire which has a characteristic strength of 250 N/mm².

For each type, a range of sizes is available, each denoted by a number which is the area of the main wires in square millimetres for a metre width. For example:

A252 is a square mesh fabric, cross-sectional area of the main wires = 252 mm² per metre.

12.12 Beam reinforcement

Figure 12.4(a) shows a reinforced concrete beam spanning between two brick walls. The beam is simply supported and may carry a deck or floor slab as a uniformly distributed load (UDL). The bending moments and

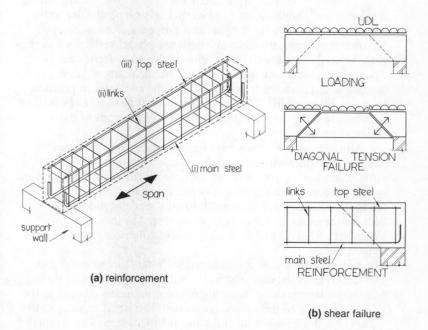

(a) reinforcement

(b) shear failure

Fig. 12.4

shearing forces vary along the beam. For a UDL, the bending moment is a maximum at mid-span and zero at the support. The shear force is zero at mid-span and a maximum at the support. Shear failure in a plain (unreinforced) concrete beam would take the form of diagonal tension cracking (b).

The assembled reinforcement for a beam is called a *cage*. Three items of reinforcement have been used in the cage shown (a).

(i) The bending moments are carried as complementary fibre stresses parallel to the axis of the beam: compression stress in the top and tension stress in the bottom. The tension stress will far exceed that which the concrete can carry unaided and so reinforcement is introduced to give the necessary tensile strength. It is the principal reinforcement in the beam and is referred to as *main steel*. Main steel is highly stressed and so high-yield 'T' bars are used, normally in the range T12–T32. The amount of main steel required is determined from structural calculations. In moderately loaded beams, the compression stress in the top of the beam is carried by the concrete.

(ii) The shear failure shown in Fig. 12.4(b) is prevented by fixing reinforcing bars across the potential lines of diagonal tension cracking. Shear reinforcement in beams normally takes the form of closed rectangular hoops bent from lengths of straight bar (Fig. 12.5(a)). The bent bar is called a *link* or *stirrup*. Ideally, shear reinforcement should be inclined so that it is perpendicular to the line of the crack, but for the practical purposes of assembling a reinforcement cage, the links in beams are placed vertically and they work in conjunction with the longitudinal steel to resist shear. Theoretically, links are not required if the shear stress due to loading is less than the shear strength of the concrete. In practice, they are still provided. They are referred to as *nominal links* and the sizes and spacings are calculated from the dimensions of the concrete profile.

 All steels may be safely bent once but only mild steel can be bent several times without fatigue or loss of strength. This makes it particularly useful when bars which have already been bent to shape by the reinforcement supplier may be further bent as part of the fixing operation. Shear links are normally made of mild steel because they may be prised open or distorted as the cages are assembled. R8, R10 and R12 are the most commonly used bars for links in beams.

(iii) The reinforcement cage is completed with bars placed in the top. *Top steel* frames the cage and works with the links to resist shear. For simple beams there are no specific requirements relating to the size and spacing of these bars, however, it is normal practice to use T bars of a size between the links and the main steel. Top steel is frequently T12, 16 or 20 mm bar. In heavily loaded beams, the top

Fig. 12.5 (a)

(b)

steel may be required to supplement the natural compressive strength of the concrete. Structural calculations are required to design this *compression steel*.

Figure 12.5 shows (a) details of a beam reinforcement cage prior to the completion of the shutter and (b) slab reinforcement at the early stages of fixing.

12.13 Slab reinforcement

Figure 12.6(a) shows a simple reinforced concrete slab spanning between two brick walls. It is simply supported and is reinforced in the bottom (tensile zone) with two sets of bars fixed in evenly spaced rows at right angles.

The *main steel* is placed parallel to the direction of span and carries the bending moments due to the weight of the slab and imposed loading. Bending moments and other loading effects vary within a structure. The design calculations for each element start with the maximum critical areas but it is not necessary to use the resulting designed reinforcement throughout. For example, as the bending moments in a simply supported slab reduce towards the supports, the area of reinforcement required also reduces. Economy of reinforcement is achieved by stopping-off or *curtailing* bars. The codes give rules for curtailment in the different elements of structure. For a simply supported slab, BS8110 normally requires that at least 50 per cent of the mid-span tension reinforcement (main steel) extends beyond the centre-line of the support by 12 times the bar size [12ϕ] (b). The remainder may be curtailed short of the centre-line of the support by $0.1 \times$ span (c). The main steel is shown as L-shaped bars fixed with the bend alternately on the left- and right-hand supports. These bars are described as 'alternately reversed'. On the reinforcement drawings, this is indicated by the letters AR after the bar identification. The side lengths of the L are determined so that the 12ϕ requirement is satisfied at the fully supported end and the curtailed end stops in the correct position. The main steel is fixed as the bottom layer where it has the greatest effect.

Although the design may assume the slab to be simply supported (i.e. 'free'), the actual construction detail used may cause some fixity and result in a negative moment at the support. Top steel may be required to prevent local cracking of the concrete, details are given in BS8110.

The reinforcement placed at right angles to the main bars is called secondary or *distribution steel* and it is used to control the effects of temperature and shrinkage, and to spread the effects of concentrated loads over a wider area. The amount of distribution steel is usually calculated as a percentage of the cross-sectional area of the concrete but may be calculated when heavy point loads are carried. In the figure, these bars are shown with a very short bend or 'bob' at one end and they

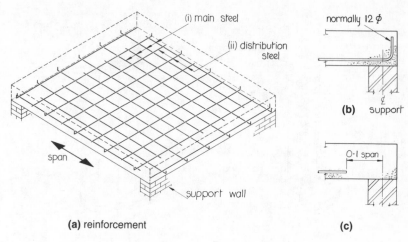

Fig. 12.6

are alternately reversed. The bob is not essential, but is useful to the steel-fixer in positioning the bars against the side shutters.

The assembly of main and distribution steel is referred to as a *mat*. The natural shear strength of the concrete is usually sufficient to carry the stresses in a moderately loaded, simply supported slab, and unlike beams, there is no requirement for nominal shear reinforcement.

All slab steel is normally T-bars, in the range T10–T25. When slabs are supported on three or four sides, the loadings are carried as bending moments in both directions. The secondary steel then becomes main steel in a load-sharing system and must be designed.

12.14 Column reinforcement

Columns carry mainly compressive forces for which the concrete is ideally suited. Two principal items of reinforcement are used in column cages (Fig. 12.7(a)).

(i) The longitudinal bars are the main steel. Columns, like other elements in a structure, must contain a minimum area of reinforcement, and even when the applied loads can be safely carried by the concrete alone, a small amount of longitudinal steel is still required. If the applied compressive forces exceed the natural strength of the concrete, longitudinal reinforcement is designed to take part of the load. Designed longitudinal steel is also used when the loading includes bending moments which cause tension in part of the column. High-yield bars are normally used for column main steel.

(ii) Columns also contain links (or stirrups) which are normally closed

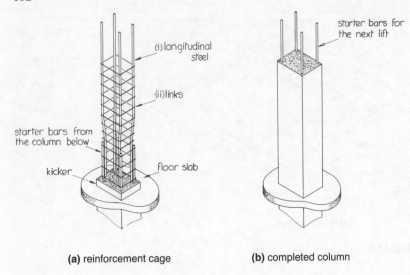

(a) reinforcement cage   **(b)** completed column

Fig. 12.7

rectangles. Their primary purpose is to contain the main steel when acting in compression and prevent it from bursting through the sides of the column. The links are normally made from mild steel bar and the size and spacing is based on the size of the main steel.

Practical limitations on the order of construction and the size of concrete pours means that structures are built as a series of linked elements. In a multi-storey building, for example, the natural order of construction is to build floors and columns alternately. The column reinforcement shown is for one storey and the full-height column is cast in one *lift*. To achieve continuity, the reinforcement cage is erected on *starter bars* which project from the column below. In turn, the main steel at this level is detailed to project from the top of the completed column to form starter bars for the next floor. The raised plinth is called a *kicker*. It is about 75 mm high and the same shape as the column. Kickers are cast directly onto the slab before the reinforcement cage is erected. They are used to locate the column shutter.

Figure 12.8 shows assembled reinforcement for (a) a column, and (b) a wall.

12.15 Wall reinforcement

Concrete walls in buildings carry mainly axial loads. They are usually reinforced with the smaller sizes of high-yield (T) bar fixed as mats of vertical main steel and horizontal secondary steel (Fig. 12.9(a)).

103

Fig. 12.8 (a)

(b)

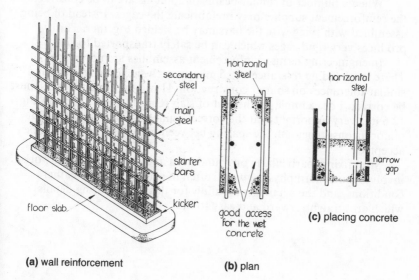

(a) wall reinforcement

(b) plan

(c) placing concrete

Fig. 12.9

secondary steel

main steel

starter bars

kicker

floor slab

horizontal steel

good access for the wet concrete

horizontal steel

narrow gap

It is theoretically possible to design concrete walls to be very slender but it is recommended that, for practical reasons, storey-height walls are made at least 200 mm thick.

The relative positions of the vertical and horizontal steel require consideration. Putting the horizontal steel on the outside of the mats (b) keeps the centre of the wall open, giving good access for placing and compacting the concrete. However, the horizontal bars are then close to the shutter and the resulting continuous narrow gap may act as a mask to the free flow of the concrete (c). On site, particular attention must be given to placing and compacting the concrete around the horizontal steel. For thicker walls, which have better core access, the detailer may prefer to put the vertical steel on the outside of the mat. When a wall carries bending moments, this may be desirable for structural reasons.

12.16 Steel fixing

The reinforcement shown in the typical details is prepared by a reinforcement supplier. It is cut from 12 or 18 m long straight bars, bent to shape in a mechanical bending machine and supplied to site ready for fixing. On site, a steel-fixer connects the bars together using soft-iron or stainless-steel tying wire. The work is normally done *in situ* although it is possible to make up reinforcement sub-assemblies in a workshop and transport and crane them into the required positions. Fixed in the shutters, the cages and mats must be completely stable and it may be necessary to include additional bars to achieve this.

When a number of similar beams or columns are to be constructed, the reinforcement supplier may prefabricate the cages. Instead of being assembled with tying wire the bars may be welded together. This produces very rigid cages which can be safely transported to site.

In engineering terms, reinforcement assemblies are not precise. There is a bending tolerance of ±5 mm on each side of a bent bar and similar tolerances on shutter construction. However, certain details must be correct. For example, in the mat of slab reinforcement shown in Fig. 12.6 it is very important that the correct number of bars are used, but minor inconsistencies in the spacing between individual bars would be accepted.

Steel fixing, particularly on foundation and civil engineering works, involves much hard physical work, often in extremely bad weather conditions, and the detailer must allow for this by preparing details which are straightforward and easy to fix.

Section 13

Areas and weights of reinforcement

13.1 Introduction

The design of a reinforced concrete structure begins with the selection of likely sizes for the beams, columns, slabs and so on. These are chosen from a combination of design code recommendations and the designer's experience.

Calculations are then carried out to check the strength of the proposed concrete sections, and to determine the amounts of reinforcement required. The calculations determine the cross-sectional area of the steel required.

Suitable reinforcement is then chosen with the aid of a bar chart, for separate bars, or fabric tables if fabric reinforcement is to be used.

13.2 Bar chart

A bar chart is a multiplication table for the areas and weights of the standard sized bars. For bar sizes from 8 to 40 mm, the area tables list:

(a) the total cross-sectional area of any number of bars from 1 to 10.
(b) the total cross-sectional area of bars at a range of spacings calculated per metre width.

Weight tables list:

(c) the total weight per metre of any number of bars from 1 to 10.
(d) the total weight per square metre for bars at various spacings.

Sectional areas of groups of bars (mm²)

bar size mm	Number of bars										bar size mm
	1	2	3	4	5	6	7	8	9	10	
6*	28	57	85	113	142	170	198	226	255	283	6*
8	50	101	151	201	251	302	352	402	453	503	8
10	78	157	235	314	392	471	549	628	706	785	10
12	113	226	339	452	565	678	792	904	1 018	1 131	12
16	201	402	603	804	1 005	1 207	1 408	1 609	1 810	2 011	16
20	314	628	943	1 257	1 571	1 885	2 199	2 514	2 828	3 142	20
25	491	982	1 473	1 964	2 454	2 945	3 436	3 927	4 418	4 909	25
32	804	1 608	2 413	3 217	4 021	4 825	5 629	6 434	7 238	8 042	32
40	1 257	2 513	3 770	5 026	6 283	7 540	8 796	10 053	11 309	12 566	40
50*	1 960	3 930	5 890	7 850	9 820	11 800	13 700	15 700	17 700	19 600	50*

Sectional areas per metre width for various bar spacings (mm²)

bar size mm	Spacings — mm									bar size mm
	50	75	100	125	150	175	200	250	300	
6*	566	377	283	226	189	162	142	113	94	6*
8	1 006	670	503	402	335	287	251	201	167	8
10	1 570	1 046	785	628	523	448	392	314	261	10
12	2 262	1 508	1 131	904	754	646	565	452	377	12
16	4 022	2 681	2 011	1 608	1 340	1 149	1 005	804	670	16
20	6 284	4 189	3 142	2 513	2 094	1 795	1 571	1 256	1 047	20
25	9 818	6 544	4 909	3 926	3 272	2 805	2 454	1 963	1 636	25
32		10 722	8 042	6 433	5 361	4 595	4 021	3 216	2 680	32
40			12 566	10 052	8 377	7 180	6 283	5 026	4 188	40
50*			19 600	15 700	13 100	11 200	9 820	7 850	6 540	50*

* Denotes non-preferred sizes.

Fig. 13.1

A typical bar area chart is shown in Fig. 13.1. The non-preferred sizes 6 mm and 50 mm are included on this table.

13.3 Bars in groups

If main steel is chosen for a beam or column, the area of steel required is taken from the table listing sectional areas of groups of bars (mm²).

Example

Calculations show that a beam requires 2756 mm² of steel. Referring to the table (Fig. 13.1) this can be achieved as:

(a) 6 no. 25 mm bars = 2945 mm²
(b) 9 no. 20 mm bars = 2828 mm²

(c) 4 no. 20 mm + 2 no. 32 mm = 1257 + 1608 = 2865 mm²
(d) 6 no. 16 mm + 6 no. 20 mm = 1207 + 1885 = 3092 mm²

The cross-sectional area of all the groups of bars chosen from the table is at least 2756 mm². Clearly with the range of reinforcement sizes available it is rarely possible to achieve the exact area required.
Reinforcing steel is expensive and the area of the chosen bars should be no more than about 10 per cent over the design requirement, and even this may be difficult to achieve in certain situations.

The bars may be all the same size ((a) and (b)) or a combination of different sizes ((c) and (d)).

Another combination of bars which would fulfil the requirement is:

2 no. 40 mm and 5 no. 8 mm bars = 2513 + 251 = 2764 mm².

There is no reason in theory why this combination would not work. In practice, such extremes of bar size are not normally combined.
When bars of different diameters are used, they tend to be grouped in similar sizes, for example:

10, 12 and 16;   16, 20 and 25;   20, 25 and 32;

13.4 Bars in mats

When reinforcement is designed for a conventional slab or wall, the calculations are made for a metre width of structure, and the area of steel required is taken from the table listing sectional areas per metre width for various bar spacings (mm²).

Example
The area of steel required in a slab is 860 mm² per metre width.
Referring to the table (Fig. 13.1):

(a) 12 mm bars at 125 mm centres   = 904 mm²/metre
(b) 16 mm bars at 200 mm centres   = 1005 mm²/metre
(c) 20 mm bars at 300 mm centres   = 1047 mm²/metre
(d) 10 mm bars at 200 mm centres ⎫ alternately   = 957 mm²/metre
     12 mm bars at 200 mm centres ⎭ placed

The combinations chosen give the required area of 860 mm²/ metre. Example (d) has two different bar sizes alternately placed. The bars would be 100 mm apart.

Of the alternatives, the 12 mm bars at 125 mm centres are the obvious choice since this area is the nearest to the requirement of 860 mm². However, other factors may affect the final choice. The slab steel may have to be linked into the reinforcement in a supporting wall where the spacing of the bars is critical, resulting in one of the alternatives being more suitable.

13.5 Weights of reinforcement

The bar chart also tabulates the weights of reinforcement. The tables are shown in Fig. 13.2. The weight of reinforcing steel is taken to be 0.007 85 kg/mm² per metre run, and the tables have been calculated accordingly.

The weight of reinforcement does not significantly affect the overall weight of a structure because, although steel is more than three times as heavy as concrete, the amount used is very small.

In structural calculations, the reinforcement may be allowed for by assuming that plain (unreinforced) concrete weighs 23 kN/m³ and normally reinforced concrete weighs 24 kN/m³. If the concrete is very

Weights of groups of bars (kg per metre run)

bar size mm	Number of bars										No of metres per tonne	bar size mm
	1	2	3	4	5	6	7	8	9	10		
6*	0.222	0.444	0.666	0.888	1.110	1.332	1.554	1.776	1.998	2.220	4505	6*
8	0.395	0.790	1.185	1.580	1.975	2.370	2.765	3.160	3.555	3.950	2532	8
10	0.616	1.232	1.848	2.464	3.080	3.696	4.312	4.928	5.544	6.160	1623	10
12	0.888	1.776	2.664	3.552	4.440	5.328	6.216	7.104	7.992	8.880	1126	12
16	1.579	3.158	4.737	6.316	7.895	9.474	11.053	12.632	14.211	15.790	633	16
20	2.466	4.932	7.398	9.864	12.330	14.796	17.262	19.728	22.194	24.660	406	20
25	3.854	7.708	11.562	15.416	19.270	23.124	26.978	30.382	34.686	38.540	259	25
32	6.313	12.626	18.939	25.252	31.565	37.878	44.191	50.504	56.817	63.130	158	32
40	9.864	19.728	29.592	39.456	49.320	59.184	69.048	78.912	88.776	98.640	101	40
50*	15.413	30.826	46.239	61.652	77.065	92.478	107.891	123.304	138.717	154.130	65	50*

Weight in kg per sq metre for various bar spacings

bar size mm	Spacing of bars (millimetres)											bar size (mm)
	50	75	100	125	150	175	200	225	250	275	300	
6*	4.440	2.960	2.220	1.776	1.480	1.269	1.110	0.987	0.888	0.807	0.740	6*
8	7.900	5.267	3.950	3.160	2.633	2.257	1.975	1.755	1.580	1.436	1.317	8
10	12.320	8.213	6.160	4.928	4.107	3.520	3.080	2.738	2.464	2.240	2.053	10
12	17.760	11.840	8.880	7.140	5.920	5.074	4.440	3.947	3.552	3.229	2.960	12
16	31.580	21.053	15.790	12.632	10.527	9.023	7.895	7.018	6.316	5.742	5.263	16
20	49.320	32.880	24.660	19.728	16.440	14.091	12.330	10.960	9.864	8.967	8.220	20
25	77.080	51.387	38.540	30.832	25.693	22.023	19.270	17.129	15.416	14.015	12.847	25
32		84.173	63.130	50.504	42.087	36.074	31.565	28.058	25.252	22.956	21.043	32
40		131.520	98.640	78.912	65.760	56.366	49.320	43.840	39.456	35.869	32.880	40
50*		205.507	154.130	123.304	102.753	88.074	77.065	68.502	61.652	56.047	51.377	50*

* Denotes non-preferred sizes.

Fig. 13.2

heavily reinforced, an appropriate increased weight can be calculated. The main use of the table is to calculate the weight of reinforcement in a structure for cost estimates and bills of quantities.

At an early stage in a design a calculation is made of the approximate weight of reinforcement in the structure. This is called a *reinforcement* or *steel estimate* and it is important for the preliminary costings of a project. As the design proceeds, more accurate estimates can be made. When the final calculations are complete and the structure has been fully detailed, the actual weight of reinforcement is obtained by adding the total lengths of each bar size used and converting to weights using the tables. For estimates and bills of quantities, the reinforcement is measured by weight in tonnes. The supply and fixing costs for the smaller bars are normally greater than for the larger, and bills itemise this accordingly.

Detailing should always be kept simple and allow for the practical problems of fixing reinforcement. For example, 40 mm bars weigh nearly 10 kg/metre, and on site steel-fixers may have to manhandle individual bars weighing 50 to 100 kg, often with difficult access. Clearly reinforcement details should be kept as simple as possible.

13.6 Fabric charts

Reinforcing fabric is made by welding cold-drawn high-tensile wire or cold-worked reinforcing bars to form rectangular grids or mats. Four

Fabric reinforcement to BS4483 (preferred fabric)

	British Standard Reference	Mesh Size Nominal Pitch of Wires		Size of Wires		Cross Sectional Area per Metre Width		Nominal Mass per Square Metre
		Main mm	Cross mm	Main mm	Cross mm	Main mm²	Cross mm²	kg
Square	A393	200	200	10	10	393	393	6.16
Mesh	A252	200	200	8	8	252	252	3.95
Fabric	A193	200	200	7	7	193	193	3.02
	A142	200	200	6	6	142	142	2.22
	A 98	200	200	5	5	98	98	1.54
Structural	B1131	100	200	12	8	1131	252	10.90
Fabric	B785	100	200	10	8	785	252	8.14
	B503	100	200	8	8	503	252	5.93
	B385	100	200	7	7	385	193	4.53
	B283	100	200	6	7	283	193	3.73
	B196	100	200	5	7	196	193	3.05
Long	C785	100	400	10	6	785	70.8	6.72
Mesh	C636	100	400	9	6	636	70.8	5.55
Fabric	C503	100	400	8	5	503	49.0	4.34
	C385	100	400	7	5	385	49.0	3.41
	C283	100	400	6	5	283	49.0	2.61
Wrapping	D49	100	100	2.5	2.5	49.0	49.0	0.76
Fabric	D98	200	200	5	5	98	98	1.54

Stock size sheets 4·8m long × 2·4m wide

Fig. 13.3

patterns of fabric are made. They are identified by a letter, A, B, C or D, and a number. The number is the area of the main reinforcement in square millimetres per metre width.

The full range of preferred fabrics is shown on the chart (Fig. 13.3) which is taken from BS4483:1969. Suitable applications for the different types of fabric are described in section 12.

There is not the same range of choice of reinforcement areas with fabrics as can be achieved by fixing separate bars. Fabric is most suited to slabs and walls where large areas of uniform reinforcement are required, and where there are few discontinuities such as holes. Fabric can be used in conjunction with individual bars to form edge details.

13.7 Construction

Subsequent sections explain how the chosen reinforcement is detailed on the drawings, how schedules are made and used in the supply of the steel, and how the drawings and schedules are used in the fixing of the steel.

Section 14

Cover to reinforcement

14.1 Introduction

When reinforced concrete structures are detailed the reinforcement must be located within the finished concrete profile. The steel must be completely surrounded by the concrete and the distance between the outermost bars and the concrete face is termed the *cover* (Fig. 14.1). Cover allows for bond (or grip) to be developed between the steel and concrete so that the structure will work properly. It also provides protection against corrosion (rusting) and fire (causing loss of strength).

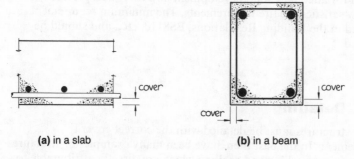

(a) in a slab   (b) in a beam

Fig. 14.1

14.2 Bond and corrosion protection

For the bond to be effective the concrete cover must be of a certain nominal thickness—this is usually taken to be at least equal to the size of the main bars.

Limit state codes allow bars to be fixed in groups or bundles. For bundles of three or more bars, the effective diameter is taken to be the size of a single bar of equivalent area, with a corresponding allowance for cover.

When reinforced concrete structures are loaded they deflect, and hairline cracks form in the (reinforced) tension areas of the concrete. Under certain circumstances moisture can penetrate these cracks and if it reaches the steel will cause rusting. The rusting causes a loss of the effective cross-section of the bars, and a consequent loss of strength. Also, the rusting steel expands to many times its former volume, and the forces which cause that expansion are considerable. Concrete is a very brittle material, and it can happen that the expanding rust causes pressures the surrounding concrete is unable to contain, and it bursts away. This is known as *spalling*. Galvanising reinforcement as a protection against corrosion has never been common practice. Stainless steel reinforcement is manufactured. It is more expensive than ordinary bar. Fusion-bonded epoxy-coated reinforcement is also available, the epoxy acts as a protective coating on the steel.

14.3 Fire resistance

Adequate cover is also required to enable a structure to resist damage during a fire. The fire resistance is measured as the amount of time in hours for which the structure would have to remain safe. Building finishes such as plaster improve fire resistance, and the concrete cover required for adequate bond development and corrosion protection normally satisfies the fire requirements. The minimum requirements are tabulated in the Building Regulations, BS8110, etc., and should be consulted.

14.4 Detailing

Clearly, structures must be detailed with the correct cover to reinforcement. In the past there have been many examples of structures suffering from cracking and spalling which are directly attributable to incorrect detailing or site fixing of reinforcement. Repairs to such structures are both difficult and expensive and can be avoided if sufficient attention is given to cover during design and construction.

14.5   Cover requirements

The amount of cover required in any particular circumstance is also
dependent on:

(a) the grade (strength) of the concrete, and
(b) the location of the concrete and its degree of exposure to corrosive
     conditions. This will vary from 'mild', where the structure would be
     completely protected from the weather, for example, floor slabs in
     an office block, to 'extreme' as in exposed marine structures.

The full range of exposure conditions are described in Fig. 14.2 which is
taken from Table 3.2 of BS8110, and appropriate reinforcement covers
are tabulated in Fig. 14.3 (Table 3.4). These requirements are for
durability. Cover needed for fire resistance is also given in the standard.
    It will be seen that:

(a) as the concrete strength increases, the amount of cover required
     decreases, and

Table 3.2 Exposure conditions	
Environment	**Exposure conditions**
Mild	Concrete surfaces protected against weather or aggressive conditions
Moderate	Concrete surfaces sheltered from severe rain or freezing whilst wet
	Concrete subject to condensation
	Concrete surfaces continuously under water
	Concrete in contact with non-aggressive soil (see class 1 of table 6.1)
	NOTE. For aggressive soil conditions see 6.2.3.3.
Severe	Concrete surfaces exposed to severe rain, alternate wetting and drying or occasional freezing or severe condensation
Very severe	Concrete surfaces exposed to sea water spray, de-icing salts (directly or indirectly), corrosive fumes or severe freezing conditions whilst wet
Extreme	Concrete surfaces exposed to abrasive action, e.g. sea water carrying solids or flowing water with pH ≤ 4.5 or machinery or vehicles

Fig. 14.2

Table 3.4 Nominal cover to all reinforcement (including links) to meet durability requirements (see note)

Conditions of exposure (see 3.3.4)	Nominal cover				
	mm	mm	mm	mm	mm
Mild	25	20	20*	20*	20*
Moderate	–	35	30	25	20
Severe	–	–	40	30	25
Very severe	–	–	50†	40†	30
Extreme	–	–	–	60†	50
Maximum free water/ cement ratio	0.65	0.60	0.55	0.50	0.45
Minimum cement content (kg/m^3)	275	300	325	350	400
Lowest grade of concrete	C30	C35	C40	C45	C50

*These covers may be reduced to 15 mm provided that the nominal maximum size of aggregate does not exceed 15 mm.

†Where concrete is subject to freezing whilst wet, air-entrainment should be used (see **3.3.4.2**).

NOTE 1. This table relates to normal-weight aggregate of 20 mm nominal maximum size.

NOTE 2. For concrete used in foundations to low rise construction, see **6.2.4.1**).

Fig. 14.3

(b) as the degree of exposure increases, the amount of cover also has to increase.

Air entrainment, which is referred to in Fig. 14.3 uses an additive in the wet concrete which draws air into the mix as small bubbles. The imported air improves the workability of the wet concrete and the durability and frost resistance of the hardened concrete. The tabulated covers are to *all* steel, including links, etc. In addition, cover should not be less than:

(i) the main bar size;
(ii) the maximum aggregate size (also see footnote * in Fig. 14.3).

For concrete cast:

(iii) against an earth face, cover should generally be not less than 75 mm;
(iv) against a blinding concrete, cover should generally be not less than 40 mm.

When concrete is to receive a surface treatment such as acid-etching or bush hammering, an additional allowance should be made for the material which will be lost.

Nominal cover requirements for highway structures designed to BS5400 are given in similar tables in part 4 of that Standard.

For water-retaining structures designed to BS8007 the cover requirement is for not less than 40 mm to all reinforcement and this may need to be increased under certain circumstances.

14.6 Dimensional tolerances

The tables refer to 'nominal' cover. Dimensional tolerances have to be allowed at several stages in reinforced concrete detailing: on the bending of reinforcing bars, the construction of the shutters, and on the fixing of reinforcement within the shutters.

There are absolute minimum values necessary for reinforcement cover. The nominal covers listed in the tables allow for a negative fixing tolerance of 5 mm, which means that the minimum cover to the steel in the finished structure (Fig. 14.4(a)) may be up to 5 mm less than the nominal values (b).

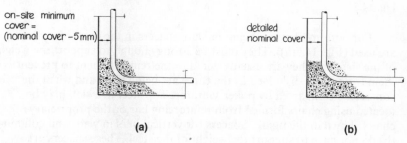

on-site minimum
cover =
(nominal cover −5mm)

detailed
nominal cover

(a)   (b)

Fig. 14.4

14.7 Drawing notes

Suitable notes for inclusion on the reinforcement drawings are:

Cover to reinforcement to be 30 mm
or
Cover to reinforcement shall be:
(a) for exposed concrete faces—40 mm
(b) for concrete against earth faces—75 mm

The figures for cover stated on the drawings are the nominal values stated in the tables. The fixing tolerance is a matter for site control by the resident engineer.

14.8 Spacers

When reinforcing steel is fixed, the required concrete cover is achieved using spacers which are placed between the bars and the shutter. They are cast into the concrete. Spacers are produced by a number of manufacturers with different types for different situations. They are normally plastic or concrete.

| (a) | (b) |

Fig. 14.5

For bottom bars in a beam or slab, spacers in the form of a chair are used (Fig. 14.5(a)). They must be strong enough to support the weight of the steel and they are usually wired to the reinforcement to prevent them from being dislodged as the cages are being fixed and when the concrete is compacted by poker-vibrator. Top steel in a slab may be located using chairs formed from reinforcing bar or the proprietary chairs shown in the figure. Spacers for vertical bars in walls and columns (b) do not have to support the weight of the steel. These spacers are clipped onto the bars and held by friction. Concrete spacers may also be used by wiring them to the reinforcement. Spacers and chairs are not detailed on the drawings. The type and location in the shutter would be proposed by the contractor and agreed by the resident engineer during the construction of the works.

Section 15

Detailing reinforcement

15.1 Introduction

Design calculations determine the principal reinforcement needed in a structure, for example, the main bars in a beam or column. Supplementary steel is also required to distribute the effects of loading, to control shrinkage and cracking, and to tie the reinforcement together to form uniform cages and mats.

Both principal and supplementary reinforcement is cut and bent to shapes given in BS4466: Specification for bending dimensions and scheduling of reinforcement for concrete. This Standard describes the bar shapes and the method of dimensioning individual reinforcing bars and fabrics.

15.2 Standard bar shapes

The standard shapes for the bending of reinforcing bars are given numbers called *shape codes*, and these are listed in tables 4 and 5 of the Standard. This information is also found in the shape code charts published and distributed by a number of reinforcement suppliers. A typical example is shown in Figs 15.1 and 15.2. Two ranges of shapes are used:

(a) 'Preferred shapes' based on table 4 of the Standard and shown in Fig. 15.1. Reinforcement should be detailed using these shapes if possible.

118

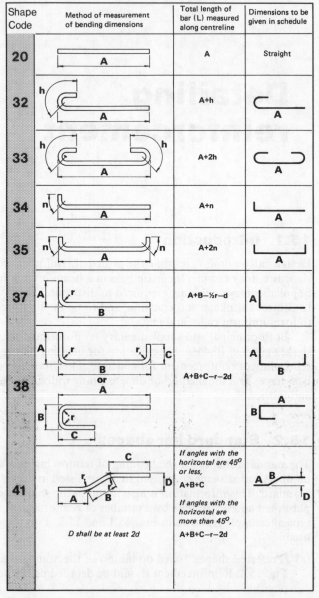

Fig. 15.1

PREFERRED SHAPES

Shape Code	Method of measurement of bending dimensions	Total length of bar (L) measured along centreline	Dimensions to be given in schedule
43		*If angle with the horizontal is 45° or less,* A+2B+C+E *If angle with the horizontal is more than 45°* A+2B+C+E −2r−4d	
51	r (non-standard)	A+B−½r−d *If r is minimum use shape code 37*	
60		2(A+B)+20d*	
62		*If angle with the horizontal is 45° or less,* A+C *If more than 45°* A+C−½r−d	
81		2A+3B+22d	
83	*Wire chairs may be substituted with the designer's approval*	A+2B+C+D −2r−4d	

120

BENDING DIMENSIONS
BS 4466

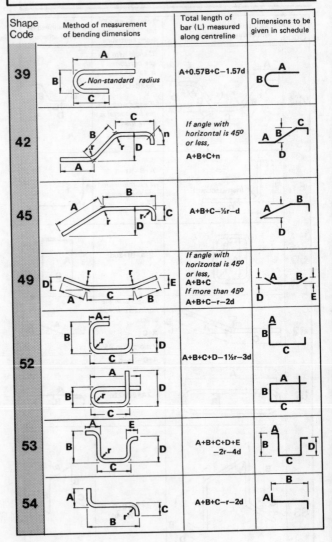

Shape Code	Method of measurement of bending dimensions	Total length of bar (L) measured along centreline	Dimensions to be given in schedule
39		$A+0.57B+C-1.57d$	
42		If angle with horizontal is 45^0 or less, $A+B+C+n$	
45		$A+B+C-\frac{1}{2}r-d$	
49		If angle with horizontal is 45^0 or less, $A+B+C$ If more than 45^0 $A+B+C-r-2d$	
52		$A+B+C+D-1\frac{1}{2}r-3d$	
53		$A+B+C+D+E-2r-4d$	
54		$A+B+C-r-2d$	

Fig. 15.2

OTHER SHAPES

Shape Code	Method of measurement of bending dimensions	Total length of bar (L) measured along centreline	Dimensions to be given in schedule
55		A+B+C+D+E −2r−4d	
65	R (non-standard)	A	R / A
72		2A+B+25d	
73		2A+B+C+10d	C (i.d.)
74		2A+3B+20d	
85	B Non-standard radius	A+B+0.57C+D −½r−2.57d	B
86		Where B is not greater than A/5 $\frac{C}{B} \pi (A+d)+8d$ where d is the size of bar L max is 12m	Helix A is the internal diameter (in mm) B is the pitch of helix (in mm) C is the overall height of helix (in mm)
99	All other shapes	To be calculated	A dimensional sketch of the shape shall be given in the schedule.

(b) 'Other shapes' based on table 5 of the Standard and shown in Fig. 15.2. These supplement the preferred shapes.

In both tables the first column lists the bar shape number. The second column shows the bent shape of the bar and gives the various dimensions which must be worked out by the detailer. The third column gives the formula for calculating the required cut length of reinforcement from which the specified bar (shape and dimensions) will be bent. The fourth column lists the bending dimensions which must be specified on the bar schedule described in Section 16. It is not necessary to give dimensions for every side of the bent bar. For certain bar shapes, standard end details apply, for example shape codes 33 and 35; and in all cases, if the cut length of the bar is right, the unspecified run-out dimension will automatically be correct.

15.3 Examples

Figure 15.3 shows (a) a number of high-yield and mild steel bars on site, and (b) the corresponding bar shapes.

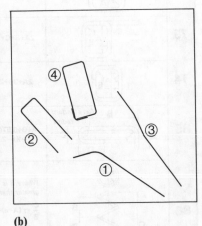

Fig. 15.3 (a)   **(b)**

① is an 'L' bar, shape code 37.
② is a 'U' bar, shape code 38.
③ is a cranked bar, shape code 41.
④ is a rectangular link, shape code 60.

When a bar is detailed, the shape code is determined from the shape of the bar regardless of the proportion of the sides. Thus all the bars shown in Fig. 15.4 are of shape code 38. The relative dimensions of *A, B* and *C* do not affect the shape code.

15.4 Non-standard shapes

If a bar is required which does not exactly match any of the standard shapes, a special shape code 99 is used. A code 99 bar can be of any form provided it is similar to the standard shapes.

15.5 Bar lengths

When a bar is detailed to turn a right angle, it is normal to show this on the reinforcement drawing as a sharp corner. In fact it is neither practical nor advisable to bend the bars this way and the reinforcement supplier shapes the bars in a bending machine which bends the steel around a circular former.

The internal radius of the bend (*r*) is the minimum possible without causing a risk of fatigue.

The overall length of a square corner on a bar (Fig. 15.5(a)) is greater than the length of a curved corner (b). When the total (cut) length is worked out, simply adding up the overall lengths of the individual sides gives an overestimate. The shape code chart allows for this by making deductions based on the number of bends (c). This explains the terms such as $-1\frac{1}{2}r -3d$ in the 'total length' columns of Figs 15.1 and 15.2. In many cases, omitting them would be of no consequence (except for a

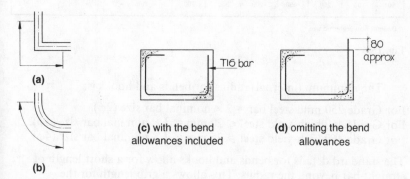

Fig. 15.4

Fig. 15.5

small waste of steel). In others, where the bar has to be an exact fit, omitting the deductions could lead to bars projecting from the concrete (d).

15.6 Bends and hooks

A number of the standard bar shapes have *n* or *h* dimensions. These are the standard bends (*n*) or hooks (*h*) and they may be required to give extra anchorage to the bar. Alternatively they may be included to help the steel-fixer locate and tie the bars together as cages and mats.

The dimensions for bends and hooks are covered by BS4466 (Fig. 15.6). They are often reproduced as part of the shape code chart. The dimensions are based on the size of the bars and the minimum radius requirements for bending bars apply.

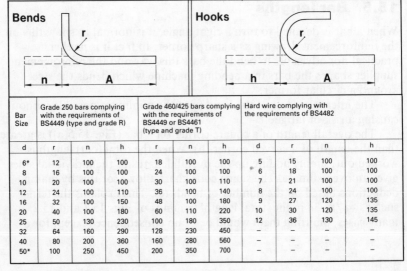

Bar Size	Grade 250 bars complying with the requirements of BS4449 (type and grade R)			Grade 460/425 bars complying with the requirements of BS4449 or BS4461 (type and grade T)			Hard wire complying with the requirements of BS4482			
d	r	n	h	r	n	h	d	r	n	h
6*	12	100	100	18	100	100	5	15	100	100
8	16	100	100	24	100	100	6	18	100	100
10	20	100	100	30	100	110	7	21	100	100
12	24	100	110	36	100	140	8	24	100	100
16	32	100	150	48	100	180	9	27	120	135
20	40	100	180	60	110	220	10	30	120	135
25	50	130	230	100	180	350	12	36	130	145
32	64	160	290	128	230	450	–	–	–	–
40	80	200	360	160	280	560	–	–	–	–
50*	100	250	450	200	350	700	–	–	–	–

* Denotes non-preferred size

Fig. 15.6

The minimum (internal) radius for bends and hooks is:

For Grade 250 mild steel bar = 2 × nominal bar size (2φ).
For Grade 460 high-yield steel ≤ 20 mm = 3 × nominal bar size (3φ).
For Grade 460 high-yield steel ≥ 25 mm = 4 × nominal bar size (4φ).

The standard details for bends and hooks allow for a short length of straight bar beyond the radius. This allows a grip length for the bar-bending machine (Fig. 15.7(a)). Although bend and hook lengths are

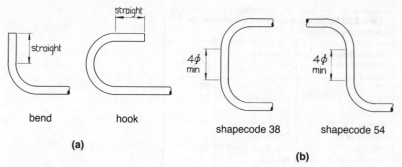

Fig. 15.7

based on bar sizes, for the smaller bars a minimum allowance of 100 mm is required.

When a bar has two or more adjacent bends, there must be a minimum straight length of 4ϕ between the bends. This requirement is shown for shape code 38 and shape code 54 bars (Fig. 15.7(b)).

15.7 Bar detailing

Bars are detailed by referring to the second column on the bar chart and selecting the appropriate shape. Then the dimensions A to E, r and n are calculated. Individual bars often have to fit between concrete faces (Fig. 15.8) and appropriate allowances have to be made to the dimensions.

Fig. 15.8

Starting from the size of the concrete structure, deductions are made for:

(a) the nominal cover to the reinforcement, at each face;
(b) a correction to accommodate the practical limits of fabrication in the cutting and bending of the bars and in the construction of the shutters.

The allowances for these limits in workmanship are shown in Fig. 15.9, which is taken from Table 3.26 of BS8110:Part 1:1985. The table lists reductions in the basic dimensions of the bars. There are no

Table 3.26 Bar schedule dimensions: deduction for permissible deviations		
Distance between concrete faces	Type of bar	Total deduction
m		mm
0 up to and including 1	Links and other bent bars	10
Above 1 up to and including 2	Links and other bent bars	15
Over 2	Links and other bent bars	20
Any length	Straight bars	40

Fig. 15.9

comparable increases for tolerances, since this would result in a loss of cover to the reinforcement.

The table also applies to water-retaining structures designed to BS5337 and a similar table is published in BS5400:Part 4 for bridge construction.

When bars do not have to be a close fit between concrete faces the allowances for cover and workmanship do not have to be so precise. For example, if the end distance x shown in Fig. 15.10 is not critical then an allowance of about 100 mm could be used for calculating the dimension A.

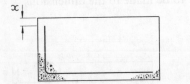

Fig. 15.10

15.8   Rounding of dimensions

BS4466 requires that the individual bar dimensions A–E should be rounded up or down to the nearest 5 mm, and that the total length of a bar should be rounded to the nearest 25 mm, for example: 1400, 1425, 1450, 1475.

15.9   Example

The use of the shape code chart is illustrated by the example shown in Fig. 15.11. The bar is assumed to be T16 with 30 mm nominal cover to

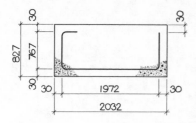

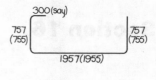

Fig. 15.11

all faces. (The dimensions of the concrete profile, 2032 and 827 mm are chosen to demonstrate the principles of reinforcement detailing. For a real structure they would probably be rounded to 2030 and 830 mm.)

The bar is a shape code 52.

Dimensioning starts from the concrete sizes. Allowances are made for the nominal cover which is 30 mm to all faces, and from these dimensions a further deduction is made for the construction tolerances in accordance with Fig. 15.9:

From the 767 mm dimension deduct 10 mm (0–1 m) = 757 mm
from the 1972 mm dimension deduct 15 mm (1–2 m) = 1957 mm

The dimensions are then rounded up or down, as appropriate, to the nearest 5 mm. The cover must not be reduced, therefore the dimensions are rounded down. Final dimensions of the bar are indicated in brackets (Fig. 15.11).

The total length of the bar is also taken from the bar chart.

$$\text{Total length} = A + B + C + D - 1\frac{1}{2}r - 3d$$

For a T16 bar, the minimum bend radius r is $3d$.

$$\begin{aligned}\text{Total length} &= 300 + 755 + 1955 + 755 - (1\frac{1}{2} \times 3 \times 16) - (3 \times 16)\\ &= 3645 \text{ mm}\end{aligned}$$

The overall length is rounded up or down to the nearest 25 mm. The cover must not be reduced, therefore the length is rounded down, so 3645 mm is adjusted to 3625 mm which becomes the cut length of the bar.

15.10 Fabric detailing

Fabric reinforcement is normally used as flat sheets but it can be bent to a number of standard shapes. The allowances for cover and tolerances already described for bars must also be included for fabric. Bending details for the fabric must also include the location of bends in relation to the transverse (cross) and longitudinal (main) wires.

Section 16

Bar and fabric schedules

16.1 Introduction

Bar and fabric schedules are listings of the reinforcing bars and fabrics shown on reinforcement drawings.

This section describes the standard format for these schedules and explains how they should be completed. Listing bars and fabric is referred to as 'scheduling'.

16.2 Bar schedules

Figure 16.1 shows a standard layout for a bar schedule based on the recommendations of BS4466:1981: Bending dimensions and scheduling of reinforcement for concrete. This would be printed as an A4 sheet.

16.3 Scheduling

The method of scheduling is explained using an example (Fig. 16.2) which details reinforcement in beams and columns.

Details of a completed bar schedule for the required reinforcement are shown in Fig. 16.3. On the schedule (Fig. 16.1), the title block includes:

(a) the name of the consultant or authority (pre-printed on the sheet)
(b) project—this should include the name of the project and the particular reinforcement drawing being scheduled.

| BSD Partnership | | | | | | | | Drawing No. | | | | |

BSD Partnership						Drawing No.				
						Bar schedule		of		
Project						Date		Rev.		
						Prep.		Ckd.		

Member	Bar Mark	Type and size	No. of mbrs	No. of bars in each	Total no.	Length of each bar † mm	Shape code	A* mm	B* mm	C* mm	D* mm	E/r* mm

This schedule complies with the requirements of BS 4466

∗ Specified in multiples of 5 mm.   † Specified in multiples of 25 mm.

Fig. 16.1

(c) drawing no.—the number of the drawing being scheduled.
(d) bar schedule—the number of the individual schedule, i.e. 2 of 6, 3 of 6. Most reinforcement drawings require several schedules.
(e) date—the date of completion of the schedule.
(f) Prep.—initials of the detailer who prepared the schedule.

The remainder of the schedule consists of columns which are completed as follows.

Member

This column is used for a sub-heading which identifies the particular member being scheduled.

Under the heading 'ground floor beams' the bars for the 5 No. beams are scheduled. It is then advisable to leave a gap. This makes a natural break and allows for extra bars to be added should they be required due to modifications to, or mistakes in, the original schedule.

Then follows the sub-heading 'ground floor columns' and all bars in the 7 No. columns are scheduled.

Bar mark

Bars are identified by their size and shape. For each reinforcement drawing, each different bar is given a unique number starting at bar mark 01, then 02, 03, 04, ..., continuing until all the bars have been defined.

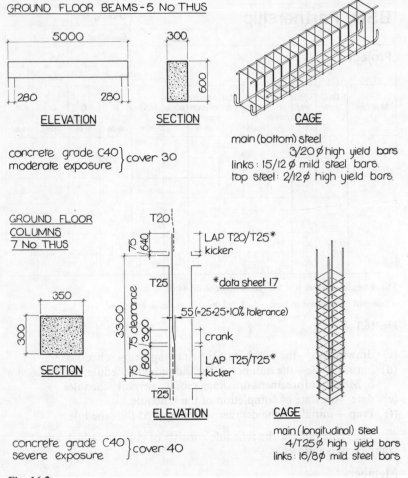

GROUND FLOOR BEAMS - 5 No THUS

5000

280   280

ELEVATION

300

600

SECTION

CAGE

concrete grade C40 } cover 30
moderate exposure

main (bottom) steel
3/20 ⌀ high yield bars
links : 15/12 ⌀ mild steel bars.
top steel : 2/12⌀ high yield bars.

GROUND FLOOR
COLUMNS
7 No THUS

350

300

SECTION

concrete grade C40 } cover 40
severe exposure

T20

75 / 640

T25

75 clearance
800 / 300
75

3300

T25

ELEVATION

LAP T20/T25*
kicker

*data sheet 17

55 (=25+25+10% tolerance)

crank

LAP T25/T25*
kicker

CAGE

main (longitudinal) steel
4/T25⌀ high yield bars
links : 16/8⌀ mild steel bars

Fig. 16.2

In a sequence of drawings, the numbering system starts again on the next drawing at bar mark 01. For example:
Dwg. No. 4206/7 bars numbered consecutively 01, 02, 03, ...
Dwg. No. 4206/8 bars numbered consecutively 01, 02, 03, ...

Type and size

This column identifies the type of reinforcement, either mild or high-yield steel, and the nominal size in millimetres.

The type is identified by the letter R, T or X (section 12.10).

The high-yield steel may be either hot-rolled or cold-worked. This is defined in the project specification or on the drawings—not on the bar schedules.

Member	Bar Mark	Type and size	No. of mbrs	No. of bars in each	Total no.	Length of each bar † mm	Shape code	A* mm	B* mm	C* mm	D* mm	E/r* mm
grd floor beams 5 No.	O1	T20	5	2	10	5150	35	4920				
	O2	T20	5	1	5	4000	20	STRAIGHT				
	O3	R12	5	15	75	1675	60	505	205			
	O4	T12	5	2	10	4900	20	STRAIGHT				
grd floor columns 7 No.	O5	T25	7	4	28	3950	41	815	300		55	
	O6	R8	7	16	112	1025	60	240	190			

Fig. 16.3

No. of members

This states the number of members for each sub-heading. In the example, there are 5 beams and 7 columns.

No. of bars in each

This states how many bars of each bar mark are in each member.

Total no.

This multiplies the number of members by the number of bars in each, to give the total number.

The remaining columns define the shape and length of each bar tabulated. 'Length of each bar mm' gives the total cut-length of bar required.

Shape code

The shape of each bar in the schedule is defined by a number from the standard range described in section 15.

Columns A–D

These are completed with the dimensions appropriate to the shape code. Look at Fig. 16.3 and fill in the dimensions *A–D* as appropriate, using Figs 15.1 and 15.2. If one is not used, leave a blank on the schedule.

For practical purposes these dimensions must be rounded up or down to the nearest 5 mm.

Column E/r

This is used either for the dimension *E* or for a non-standard radius *r*,

132

for example shape code 51 (Fig. 16.4(a)). This is a specific radius larger than the minimum bend radius.

When a shape code 20 bar is detailed, the A to E columns may be left blank, since no bending dimensions are required. Alternatively, the word 'straight' can be written across them.

If a particular bar is required which is not defined by the standard shapes, then shape code 99 is used. Columns A–D, E/r are not space (Fig. 16.4(b)). The dimension that is free to take up the tolerances is put in parentheses.

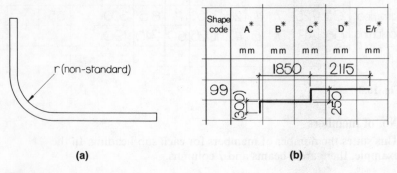

(a)   (b)

Fig. 16.4

16.4 Alternative format

An alternative format for the A–D, E/r part of the schedule uses the line drawing technique. In place of these columns, the bar schedule is blank and each bar required is drawn and dimensioned up. This system is not often used, the tabular method being preferred.

16.5 Calculating bar lengths

The total cut length for each bar is calculated by adding up the length of each part and then deducting bend allowances.

The total length of a bar has to be rounded to the nearest 25 mm. Whether the length is rounded up or down will depend on the particular situation.

16.6 Reinforcement suppliers

The term 'bar schedule' is used to describe the listing which a detailer prepares from a reinforcement drawing. The bars are listed on the schedule in the order in which they will be required during construction, regardless of type and diameter.

At the time of construction, the bar schedules are sent to a reinforcement supplier who rewrites the schedules as cropping and bending lists which tabulate the bars in descending order of cutting length and groups the bars by type and size. This is the most efficient way to coordinate the cutting and bending work. It also acts as a final check on the accuracy of the schedule. The term 'bending schedule' is often used to describe the detailer's listing—the bar schedule. Although this is strictly incorrect according to the requirements of BS4466, it is in common use.

The 'length of each bar' column on the bar schedule is particularly important since the first stage of the reinforcement supplier's operation is to cut the bars to the overall lengths given, and an error would be difficult to correct during the bending operation. Also, steel is paid for by the tonne and weights of steel are taken off the schedule by totalling: length of bar (metres) × weight per metre = weight of bar.

16.7 Fabric schedules

Fabric reinforcement must also be scheduled and the scheduling must include details of the size of the wires and mesh, the cut size of the sheets and any bending requirements. There are three categories of fabric—designated, scheduled and detailed. Each require a different method of scheduling. For certain fabric details the standard bar schedule can be used and there is also a recommended BS fabric schedule.

Designated fabric is the preferred BS type which is produced to a standard range of wire and mesh sizes. Flat sheets of designated fabric can be listed on a standard bar schedule by ignoring the column headings and writing the details across the sheet (Fig. 16.5).

Member	Bar Mark	Type and size	No. of mbrs	No. of bars in each	Total no.	Length of each bar † mm	Shape code	A* mm	B* mm	C* mm	D* mm	E/r* mm
	5 No	SHEETS	OF	B503 MESH	4·2m LONG ×2·0m WIDE.							

Fig. 16.5

16.8 Bending fabric

Fabric can also be bent to some of the normal range of shape codes, in which case the details should be listed on a fabric schedule. The BS4466 recommended layout for a fabric schedule is shown in Fig. 16.6. It can

BSD Partnership									Drawing No.						
									Fabric schedule		of				
Project									Date		Rev.				
									Prep.		Ckd.				
Fabric mark	No. of wires	Size of wires mm	Pitch † mm	Length † mm	ovrhang O_1 O_3 mm	O_2 O_4 mm	Sheet length "L" † m	Sheet width "B" † m	No. of sheets	Special details and/or bending dimensions					
										Shape code	Bending instruction	A* mm	B* mm	C* mm	D* E/r* mm mm

This schedule complies with the requirements of BS 4466

* Specified in multiples of 5 mm   † Specified in multiples of 25 mm

Fig. 16.6

also be used for scheduled fabric. Like a bar schedule, it is printed as an
A4 sheet. Scheduled fabric is made to the designer's requirements for
wire size and spacing. Overhangs can also be included. These can be
used to tie the mesh to adjacent reinforcement. The identification
symbols used are shown in Fig. 16.7. The diagram assumes that the
sheets are always viewed with the cross-wires uppermost (inset sketch).

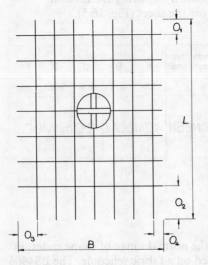

Fig. 16.7

L is the length of the longitudinal wires (which are not necessarily the longer wires on the sheet).

B is the length of the cross-wires.

O_1 and O_2 are the overhangs on the longitudinal wires.

O_3 and O_4 are the side overhangs of the cross-wires.

Detailed fabric is made to the designer's requirements and may have irregular wire and mesh sizes. A dimensional drawing is required to illustrate the details and this is best done as an A4 sketch. Reinforcement suppliers produce suitably annotated pads for this purpose.

Section 17

Joints

17.1 Introduction

The cages and mats of reinforcement used in concrete structures are assembled from individual bars which must be of manageable lengths and weights. This means there will be frequent breaks in the reinforcement, even when the design assumes that the structure is continuous.

This section explains how structural connections are made between reinforcing bars by lapping and other means.

17.2 Bond

It is essential to the working of reinforced concrete that the concrete grips the steel. To design joints between bars the strength of this grip must be determined. As newly placed concrete hardens it dries out and shrinks. If the concrete is reinforced, it shrinks onto the bars and grips them (*shrinkage*). In addition, the hardened concrete sticks to the lightly rusted steel (*adhesion*). The combination of shrinkage and adhesion is called *bond*. It is measured as a stress in N/mm^2.

The strength of bond is dependent on the surface profile of the reinforcement and the strength of the concrete mix. Mild steel bar, which is completely smooth and round, will achieve an ultimate anchorage bond strength of $1.77\,N/mm^2$ with a typical structural concrete. Deformed high-yield bars have a higher bond strength than plain bars because the raised ribs cause mechanical interference—a typical ultimate anchorage bond strength is 3.16 N/mm².

17.3 Anchorage lengths

Figure 17.1 shows a reinforcing bar which must be effectively terminated at X–X. At this point, the bar may be highly stressed and it cannot be terminated abruptly. Instead it must be locked-off so that the force P can be transferred to the surrounding concrete over a predetermined length. The bar is said to be *anchored* and the distance beyond X–X to the end is the *anchorage length*. The necessary resistance is derived from the anchorage bond between the concrete and steel, effective on the surface area of the bar over this length. To prevent failure, the strength of the anchorage must be at least equal to the force P.

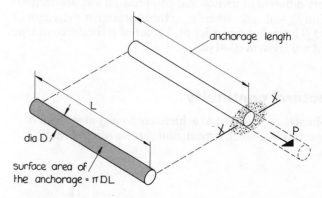

Fig. 17.1

Assume that the allowable bond stress is f_{bond} and the tension stress in the bar at X–X is f_{steel}.

The applied force P = stress in the bar × cross-sectional area

$$= f_{steel} \times \left[\frac{\pi}{4} D^2 \right]$$

The strength of the anchorage = concrete/steel bond stress × surface area of the anchorage.

$$= f_{bond} \times \pi D L$$

The strength of the anchorage must be equal to the force P,

$$f_{bond} \times \pi D L = f_{steel} \times \left[\frac{\pi}{4} D^2 \right]$$

then the anchored length $L = \left[\dfrac{f_{\text{steel}}}{4 \times f_{\text{bond}}}\right] D$

Using BS8110 derived symbols the equation becomes

$$l = \left[\dfrac{0.87 f_y}{4 f_{\text{bu}}}\right] \phi$$

For a particular concrete/steel combination, the bond strength (f_{bu}) and steel strength ($0.87 f_y$) have fixed values (it is normal to assume the steel is working at the full design strength). This means that the anchored length (l) can be simply expressed as a multiple of the bar size ϕ, for example 34ϕ, 41ϕ.

Although the bar in Fig. 17.1 is shown in tension, the same principles of anchorage also apply to bars in compression. The values of bond strength are different in tension and compression and are derived from the equation $f_{\text{bu}} = \beta \sqrt{f_{\text{cu}}}$, where f_{cu} is the characteristic strength of the concrete and β is the bond coefficient. Values of β for different types of reinforcement are given in BS8110.

17.4 Structural continuity

The principle of anchorage is used to achieve continuity at structural connections between bars. Some typical joints are shown in Fig. 17.2.

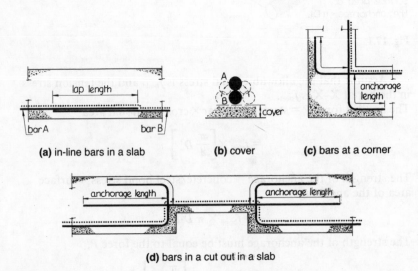

(a) in-line bars in a slab   **(b)** cover   **(c)** bars at a corner

(d) bars in a cut out in a slab

Fig. 17.2

Structural continuity does not require the bars to be in physical contact and in each case the reinforcement is effectively continuous along the line of dots.

The most common method of achieving continuity across joints is to place the bars side by side within the concrete structure (a). This is called *lapping* and the amount of overlap is called the *lap length*. The lap is used to transfer a force in the bar A into the bar B. The force passes from bar A into the surrounding concrete by bond action then across to the concrete surrounding bar B and into that bar, also by bond action. The lap length for the connection is derived from the anchorage length of the individual bars.

Bars connected in this way are often close to the concrete face and for the bond to develop properly there must be a minimum amount of concrete surrounding the steel. This affects the cover requirements (Section 14). The orientation of the bars does not affect the strength of the lap provided the minimum cover is maintained (b).

When a structural connection is made between bars at right angles, the bars must be crossed over and anchored in the concrete beyond the cross-over point (c). This detail frequently occurs in retaining wall and culvert construction. Although the bars are not side by side as in the normal lap, structural continuity is achieved because each bar is adequately bonded beyond the cross-over point, and the necessary load transfer can take place. The detail (d) would be appropriate where reinforcement must be effectively continuous across a recess in the concrete profile.

17.5 Tension and compression laps

The actual tension and compression laps required between bars are derived from the basic anchorage lengths. In certain situations the basic length may not be fully effective and must be increased. This can occur when bars are lapped close to the concrete surface or sets of bars are lapped close to one another.

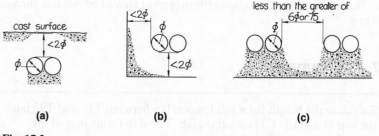

Fig. 17.3

Designed tension laps must be at least equal to the anchorage length but should be increased by a factor of 1.4 when:

(i) bars are lapped in the top of an element as cast and the minimum cover is less than twice the bar size (Fig. 17.3(a)).

(ii) bars are lapped at a corner and the minimum face or side cover is less than twice the bar size (b).

(iii) the clear distance between adjacent laps is less than the greater of 6ϕ or 75 mm (c). For this requirement the clear distance used for size 8, 10 and 12 bars would be 75 mm;

e.g. size 8: $6\phi = 6 \times 8 = 48 < 75$; use 75 mm

size 12: $6\phi = 6 \times 12 = 72 < 75$; use 75 mm

For size 16 and above the clear distance would be based on 6Φ;

e.g. size 16: $6\phi = 6 \times 16 = 96 > 75$; use 96 mm

size 40: $6\phi = 6 \times 40 = 240 > 75$; use 240 mm.

When both conditions (i) and (ii) or (i) and (iii) apply, the lap length should be increased by a factor of 2.0.

Design compression laps should be 25 per cent greater than the corresponding compression anchorage length.

Values of anchorage and lap length for a range of reinforcement types and concrete strengths are shown in Fig. 17.4, which is taken from Table 3/29 of BS8110:Part 1:1985. They are expressed as multiples of the bar size ϕ.

When bars of different sizes are lapped, the lap length may be based on the smaller bar.

17.6 Minimum laps

Bars are often lapped at a point where the stress in the steel is very low. Designing a lap or anchorage using the basic formulae produces a length which is proportionally small—perhaps 100 mm. A lap of 100 mm would be neither advisable nor practical because in-line bars could not be properly fixed to one another. In this situation a minimum lap length is required. For structures designed to BS8110, the minimum lap length for bars should be not less than 15ϕ or 300 mm, whichever is the greater. Laps using bars of different sizes may be based on the smaller bar.

The minimum lap in fabric reinforcement should be not less than 250 mm.

17.7 Examples

1. Calculate the length for a full tension lap between T16 and T25 bars in the top of a grade C35 concrete slab. The slab forms part of an internal floor in an office development.

Table 3.29 Ultimate anchorage bond lengths and lap lengths as multiples of bar size

Reinforcement type	Grade 250 plain	Grade 460			
		Plain	Deformed type 1	Deformed type 2	Fabric
Concrete cube strength 25					
Tension anchorage and lap length	39	72	51	41	31
1.4 × tension lap	55	101	71	57	44
2.0 × tension lap	78	143	101	81	62
Compression anchorage length	32	58	41	32	25
Compression lap length	39	72	51	40	31
Concrete cube strength 30					
Tension anchorage and lap length	36	66	46	37	29
1.4 × tension lap	50	92	64	52	40
2.0 × tension lap	71	131	92	74	57
Compression anchorage length	29	53	37	29	23
Compression lap length	36	66	46	37	29
Concrete cube strength 35					
Tension anchorage and lap length	33	61	43	34	27
1.4 × tension lap	46	85	60	48	37
2.0 × tension lap	66	121	85	68	53
Compression anchorage length	27	49	34	27	21
Compression lap length	33	61	43	34	27
Concrete cube strength 40					
Tension anchorage and lap length	31	57	40	32	25
1.4 × tension lap	43	80	56	45	35
2.0 × tension lap	62	113	80	64	49
Compression anchorage length	25	46	32	26	20
Compression lap length	31	57	40	32	25

NOTE. The values are rounded up to the whole number and the lengths derived from these values may differ slightly from those calculated directly for each bar or wire size.

Fig. 17.4

The basic anchorage length for these bars is 34ϕ based on the size of the smaller bar: $34\phi = 34 \times 16 = 544$ mm.

The detail should be checked against the factoring requirements for bars close to the top of a slab (Fig. 17.3a). The degree of exposure is 'mild'. The (smaller) bar size = 16 mm, $2\phi = 32$mm.

The nominal cover = 20 mm, which is less than 2ϕ and so the basic anchorage length must be factored by 1.4.

$$\text{design tension lap length} = 1.4 \times 544 = 762 \text{ mm} \quad (765)$$

This is the length used on the reinforcement detail.

---

2. Calculate the minimum compression lap between T16 and T25 bars in a grade C35 concrete slab.

The lap may be based on the smaller bar.

$$\text{minimum lap: } 15\phi = 15 \times 16 = 240 \text{ mm}$$
$$\text{or} \quad 300 \text{ mm} \qquad \text{use } 300 \text{ mm}$$

17.8 Hooks and bends

Anchorages may be required in areas of a structure where space is restricted and the bars may be hooked or bent to keep within the concrete profile. Bends have an anchorage value greater than the equivalent length of the straight bar. The effective anchorage lengths are given in the Standard (Fig. 17.5).

Anchorage lengths

For a 180° hook, the greater of:

(i) $8r$ but not greater than 24ϕ;
(ii) the actual length of the hook including the straight portion.

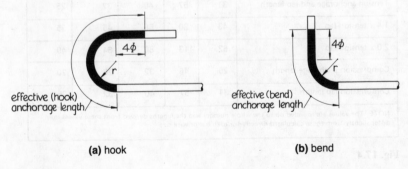

(a) hook   **(b)** bend

Fig. 17.5

For a 90° bend, the greater of:

(i) 4r but not greater than 12ϕ;
(ii) the actual length of the bend including the straight portion.

These requirements are interpreted in the table (Fig. 17.6) which shows the effective anchorage lengths of hooks and bends in multiples of the bar size ϕ.

bar size mm	Grade 250 bars minimum bend radius: all sizes – 2ø		Grade 460 bars minimum bend radius: up to size 20 – 3ø size 25 and over – 4ø		bar size mm
	bend	hook	bend	hook	
6*	19	19	20	24	6*
8	15	16	16	24	8
10	13	16	14	24	10
12	11	16	12	24	12
16	9	16	12	24	16
20	8	16	12	24	20
25	8	16	12	24	25
32	8	16	12	24	32
40	8	16	12	24	40
50*	8	16	12	24	50*

* denotes non-preferred sizes

Fig. 17.6

17.9 Laps in beams and columns

The Standard also details particular requirements for the containment of main reinforcement at laps in beams and columns when both bars are

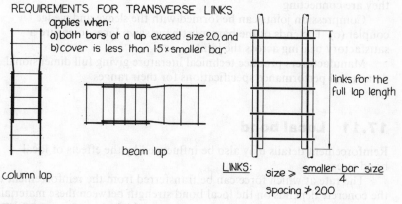

REQUIREMENTS FOR TRANSVERSE LINKS
applies when:
a) both bars at a lap exceed size 20, and
b) cover is less than 15 × smaller bar.

links for the full lap length

beam lap

column lap

LINKS: size $\geqslant$ $\dfrac{\text{smaller bar size}}{4}$

spacing $\not> 200$

Fig. 17.7

144

larger than size 20 and there is limited cover (Fig. 17.7). The detail ensures that elements are adequately strengthened in this area of potential weakness.

17.10 Couplers

Tension and compression joints can also be made between ribbed high-yield bars using purpose-made couplers to form a mechanical connection.

A simple tension joint is formed with a single sleeve which is compressed onto the reinforcing bar using a hydraulic press (Fig. 17.8(a)). Other couplers use combinations of threaded sleeves and studs (b).

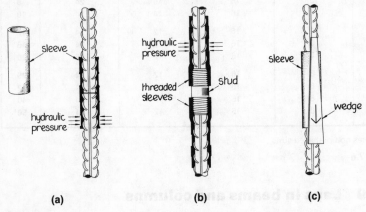

(a)   (b)   (c)

Fig. 17.8

Each must have a tensile strength at least equal to that of the bars they are connecting.

Compression joints can be formed with the sleeve and wedge coupler (c). The ends of the bars must have sawn faces to ensure a satisfactory bearing across the joint.

Manufacturers produce technical literature giving full dimensional details and performance specifications for their ranges.

17.11 Local bond

Reinforcement details may also be influenced by the effects of local bond stress.

The rate at which force can be transferred from the reinforcement to the concrete depends on the local bond strength between these materials. This can be exceeded in areas where there are high shear forces and relatively small amounts of tension reinforcement—for example, short-span beams carrying heavy loads. However, when bars are fully anchored in the ways described, the effects of load bond stress may be ignored.

17.12 Construction joints

Reinforced concrete must be cast in stages, and construction joints will be required throughout a structure. At a joint in the concrete, a break may also be required in the reinforcement, and the bars must be lapped to maintain continuity. The sequence of construction and the location of joints is the contractor's responsibility, but there are conventional positions for joints decided by the practical limitations of construction. These include the length and weight of the reinforcing bars and the overall length or height of a concrete pour. Reinforcement should be detailed assuming joints at the conventional locations and the joints should be indicated on the drawings.

17.13 Joint detailing

In prismatic elements such as beams and columns, laps in the

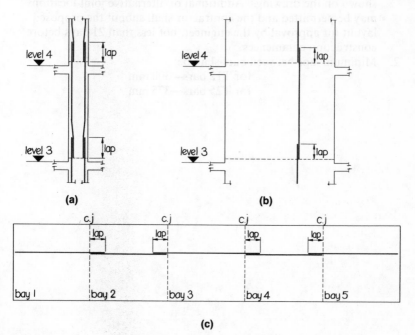

(a)   **(b)**

(c)

Fig. 17.9

longitudinal reinforcement are normally detailed with cranked, shape code 41 bars which fit tightly into the enclosing link system and maintain the correct cover. The longitudinal reinforcement in the column (Fig. 17.9(a)) is lapped in this way. The bent dimensions of the bar are determined from the floor-to-floor height, the bar sizes and the calculated lap lengths. On the detail drawings, the height of the kicker may be given, but the lap length is not normally stated since, if the bars are correctly detailed, the laps should automatically be correct.

At the laps between bars in plate-like elements such as slabs and walls, the correct cover can be maintained using straight, shape code 20 bars (b). The horizontal bars in the wall (c) are lapped at construction joints. They are also straight. Long walls are usually cast in lengths of about 5 metres and, to minimise the effects of shrinkage, an alternate bay casting sequence may be used. Bays 1, 3 and 5 are cast with projecting reinforcement, then bays 2 and 4 are cast with laps at each end. The horizontal bars in a simple wall of this type are normally required only as distribution steel and minimum laps are used. If the bays are cast progressively 1–2–3–4–5, the lap positions are detailed accordingly.

On the reinforcement detail drawings, typical 'notes box' items for construction joints would be:

1. The reinforcement is detailed to the layout of construction joints shown on the drawings. Additional or alternative joint locations may be permitted and the contractor shall submit the proposed layout for approval by the engineer, not less than 21 days before construction commences.
2. Minimum laps between bars shall be:
   for T12 bars—300 mm
   for T25 bars—375 mm

Section 18

Bar spacing

18.1 Introduction

Cages and mats of reinforcement must be detailed with suitable spacings between the bars. Bars detailed too close together may cause difficulties during construction, while bars detailed too far apart may create problems in the appearance and long-term durability of the finished structure. Guidance on suitable bar spacings is given in the design codes. Basic spacings are normally chosen as part of the structural calculations but, at the detailing stage, when the full structure is drawn and the bars are detailed for each individual element, problems may become apparent, requiring modifications to the chosen details.

This section describes the rules for spacing reinforcement.

18.2 Minimum spacings

Reinforcement must be detailed with sufficient space between the bars to allow concrete to be placed and properly compacted. The concrete must be able to flow freely to all parts of the shutter and there must be adequate access for the immersion vibrators normally used for compaction. Top steel in the narrow, rectangular beam (Fig. 18.1(a)) would require careful detailing to ensure the necessary access, and often heavily reinforced sections are detailed to include special gaps (windows) between the bars. The immersion vibrator used by the contractor would probably be 40–50 mm diameter and access windows detailed in the

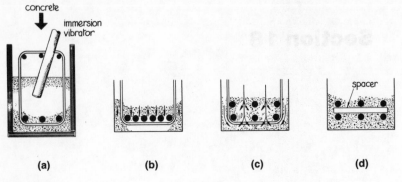

Fig. 18.1

reinforcement might be 100 × 200 mm at about 1.5–2.0 m centres. To ensure that the concrete can be placed and adequately compacted in even the most heavily reinforced sections, design codes specify minimum spacings between bars.

Consider a 250 mm wide beam reinforced with 6 no. T20 bars. If this steel is fixed as a single row (Fig. 18.1(b)) the gap between adjacent bars is only about 8 mm. The concrete is placed from above and the bars will trap the larger-sized aggregate and restrict compaction. This may result in a weak, poorly compacted concrete below the steel, and a reduction in the concrete/steel bond. By placing the bars in two rows (c) the clear horizontal space between bars is about 50 mm, allowing freer access for the concrete and effective compaction. The vertical spacing between rows is most simply achieved with spacers—short lengths of straight bar—which are scheduled as part of the reinforcement (d).

The BS8110 requirements for minimum spacings for individual bars are shown in Fig. 18.2. The horizontal distance between bars (a) should not be less than:

(i)   the maximum aggregate size (h_{agg}) + 5 mm; or
(ii) the bar size

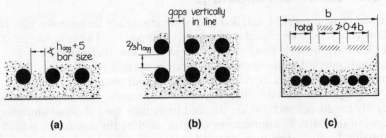

Fig. 18.2

For example, with T12 bars and concrete with 20 mm maximum aggregate the distance should not be less than:

(i) $h_{agg} + 5 = 20 + 5 = 25$ mm; or } take 25 mm
(ii) $\phi = 12$ mm

and with T32 bars and concrete with 20 mm maximum aggregate:

(i) $h_{agg} + 5 = 20 + 5 = 25$ mm or } take 32 mm
(ii) $\phi = 32$ mm

The minimum vertical distance between bars placed in rows (b) should be $\frac{2}{3}h_{agg}$ (i.e. 14 mm for a 20 mm aggregate) and the gaps between corresponding bars in each row should be vertically in line.

At laps, the minimum width taken up by the bars in a layer should not exceed 40 per cent of the width of the concrete (c). This requirement may be critical in beams which are heavily reinforced.

Similar spacing rules are given for bars in pairs, and bundled bars.

18.3 Maximum spacings

Structural concrete has a high compressive strength but the tensile strength is low (about 10 per cent of the compressive value). In reinforced-concrete structures, the tension stresses, which may be due to the effects of flexure, temperature or shrinkage, are normally carried by the reinforcement, but only after the concrete has cracked. Tension cracking cannot be prevented but it must be controlled since excessively wide cracks may allow atmospheric moisture to penetrate to the reinforcement, leading to corrosion.

Calculations can be made to determine theoretical crack widths. They are based on the stress in the bars, the bar spacings and other factors. But cracking is random and the actual widths may vary appreciably from the theoretical values. For most practical design, suitable maximum bar spacings are tabulated in BS8110. These values are conservative, but are normally acceptable unless it is felt that individual calculations are necessary. Maximum clear distances between bars are given for reinforcement of different strengths and the table allows for the effects of redistributing the bending moments which occur in continuous structures. The values are tabulated for tension reinforcement in beams. The same values may also be used for slab reinforcement, although relaxations may be applied in certain circumstances.

For simple cantilevers and simply supported beams and slabs in which there is no redistribution (Fig. 18.3) the appropriate values are:

(i)   mild steel reinforcement (grade 250)—the maximum clear distance between adjacent bars is 300 mm;

150

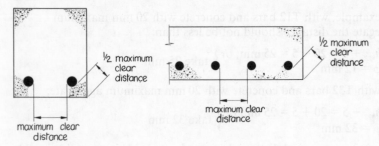

Fig. 18.3

(ii) high-yield reinforcement (grade 460)—the maximum clear distance between adjacent bars is 160 mm.

The clear distance from a corner to the nearest longitudinal bar should not exceed half these values.

These requirements may be relaxed for slabs which are lightly reinforced or of only moderate thickness.

Increases in the clear distance between bars in slabs are permitted when the percentage of tensile reinforcement is less than 1 per cent. Bar spacing checks are not normally required for slabs:

(i)   less than 200 mm thick, reinforced with high-yield steel (grade 460);
(ii) less than 250 mm thick, reinforced with mild steel (grade 250);
(iii) containing less than 0.3 per cent reinforcement (based on $100As/bd$).

Although this allows spacings to be increased, the maximum spacing used for main steel in a slab is normally 250–300 mm.

Reinforcement in beams and slabs is often detailed using bars of more than one size. This may allow the area of steel chosen from the bar chart to be as close as possible to the area required by the design. For spacing checks when different sizes are used, bars should be ignored which are less than 0.45 times the maximum size.

18.4 Example 1

Choose suitable reinforcement for the simply supported reinforced concrete beam shown in Fig. 18.4. The beam is 500×320 mm wide. The concrete is grade C40 with 20 mm maximum aggregate, and the calculated reinforcement areas are:

Main steel: 970 mm²—high-yield bar
Links: 750 mm²/m—mild steel bar
Nominal cover to the reinforcement—30 mm
Links: from the bar area chart, use R10@200°/c.

$$(2 \times 392 = 784 > 750 \quad \text{OK})$$

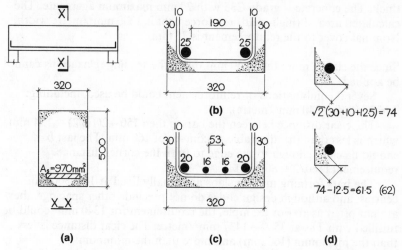

Fig. 18.4

Main steel: could be chosen from a number of combinations of bars,
including: 2 T25s (982 mm²)

2 T16 + 2 T20 (= 402 + 628 = 1030 mm²)

The 2 T25s are the first choice, since the area is closer to the design
requirement of 970 mm². Checks are made on the proposed bar layout.

The maximum permitted clear distance between bars is 160 mm.

From the sketch made for the 2 T25s (b), the clear distance between
bars is 190 mm, which exceeds the permitted value and is unacceptable.
The second choice 2 T20/2 T16 is sketched (c). Clearly with four evenly
spaced bars, the maximum clear distance will not be exceeded, however,
it is advisable to check that the bars are not then too close together. The
minimum clear distance is 53 mm, which exceeds

(i) $h_{agg} + 5 = 20 + 5 = 25$ mm $\Big\}$ OK
(ii) maximum bar size = 20 mm

The check continues with the maximum corner distance which should
not exceed 160/2 = 80 mm. For the bar layout chosen, the corner
distance is 62 mm—satisfactory.

If the gaps between the evenly spaced bars were too small the bars
could be paired or bundled to maintain the necessary free access.
Spacing checks are then required on the revised bar layout. The T16s are
within 150 mm of the link.

Finally, if laps occur in this steel, a check would be required for the
width of bars in the layer (not greater than 40% of the breadth *b*).

18.5 Example 2

Choose suitable reinforcement for a simply supported slab 280 mm

152

thick. The concrete is grade C35 with 20 mm maximum aggregate. The calculated area of (high-yield) reinforcement is 1300 mm²/metre width. Nominal cover to the reinforcement is 35 mm.

Since the slab is more than 250 mm thick, the spacing relaxations cannot be applied.

Several combinations of reinforcement could be used, including: T16@150^c/$_c$ (1340 mm²/metre).

The clear distance between the bars is then $150 - 2(16/2) = 134$ mm, which is less than the allowable maximum of 160 mm. The last bar should be close enough to the edge to fulfil the corner distance requirement ($160/2 = 80$ mm).

The bars forming mats in slabs are usually fixed at 100–300 mm centres, and although design codes do not preclude other spacings, they are not often used. For example, the requirement for 1340 mm² could be fulfilled with T8s at 35^c/$_c$—1437 mm²/metre. The clear distance is less than the maximum (160 mm) and more than the minimum:

$$\left.\begin{array}{r} h_{\text{agg}} + 5 = 25 \\ \text{max. bar size} = 8 \end{array}\right\} \text{actual distance } 35 - 8 = 27 \text{ mm} \qquad \text{OK}$$

But in practice the reinforcement would not be detailed in this way since it would be very difficult and expensive to fix so many small bars.

18.6 Side reinforcement

Reinforcement is required to regulate the effects of cracking in the side faces of deep beams.

When the overall depth of a beam exceeds 750 mm, additional

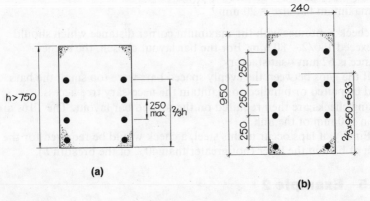

(a)

(b)

Fig. 18.5

longitudinal bars must be provided in the sides for $\frac{2}{3}$ of the overall depth from the tension face (Fig. 18.5(a)). The bar size should be not less than

$$\sqrt{\frac{s_b\, b}{f_y}}$$

where: s_b = spacing of the bars (maximum 250 mm)
b = breadth of the beam at the section considered.
A minimum size is not stipulated; the $0.45\phi_{max}$ rule does not apply.

18.7 Example 3

Determine the minimum diameter for crack control reinforcement in the sides of the beam shown in Fig. 18.5(b). Assume that high yield bars will be used at 250 mm centres.

$s_b = 250$ mm
$b = 240$ mm $\qquad \phi = \sqrt{\dfrac{250 \times 240}{460}} = 11.4$ mm
$f_y = 460$ N/mm² $\qquad\qquad\qquad$ use T12 bars.

These bars are required for $\frac{2}{3}$ of the overall depth of the beam (633 mm) but the 250 mm spacing will take them to about 800 mm. If these bars must be restricted to $\frac{2}{3}$ depth, the centres should be closed.

18.8 Other structures

The requirements for minimum and maximum bar spacings in reinforced concrete highway structures are given in BS5400:Part 4. Although the details vary, they are similar to the BS8110 requirements and are not considered separately.

The control of cracking is particularly important in liquid-retaining structures and BS8007 includes extensive provisions for the effects of: shrinkage which occurs as the cast concrete cures, temperature changes, and direct and bending stresses in the completed structure.

Section 19

Minimum and maximum reinforcement

19.1 Introduction

The calculations prepared when a structure is designed give the areas of reinforcement required in the beams, slabs, and other elements. Occasionally, the area of steel required by the design is very small. This may occur in minor structural elements or where, for practical reasons, the element is much larger than is necessary for the amount of load carried.

However, for all but the most lightly loaded structures, there is a minimum amount of steel required for each element in a reinforced structure, regardless of the requirements of the design calculations. The purpose of this *minimum* steel is to distribute loading, shrinkage and thermal effects uniformly within the element. Without this distributing effect, concentrations of stress could cause unacceptable cracking in the concrete.

Similarly, there are *maximum* amounts of steel which can be used because too much steel causes congestion and the proper placing and compacting of the concrete becomes difficult.

This section describes the minimum and maximum reinforcement requirements for structures designed to BS8110. The requirements for bridges designed to BS5400 are very similar and are not considered separately. BS8007, the code for water-retaining structures, includes particular requirements for the very thick sections often used in tanks.

Minimum and maximum reinforcement is normally calculated as a percentage area of the concrete section.

19.2 Solid slabs

Minimum reinforcement

Reinforced concrete slabs are usually made as thin as possible for a particular combination of load and span and require moderate or substantial reinforcement. Occasionally slabs are used which are thicker than is necessary for structural reasons and the design indicates that only a small amount of main reinforcement is needed. This should be checked against the requirements for minimum steel.

Distribution steel in slabs is normally chosen directly from the appropriate minimum requirements.

The areas of minimum main and distribution steel are shown for the solid slab (Fig. 19.1(a)).

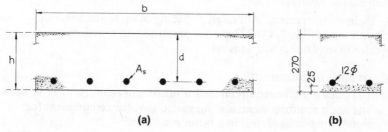

(a)   **(b)**

Fig. 19.1

h = the overall slab thickness
b = the breadth (taken to be 1000 mm to coordinate with the bar area charts).
A_c = the total area of the concrete (= bh for a solid slab)
A_s = minimum recommended area of reinforcement.
d is measured to the centre line of the reinforcement (effective depth)

Main reinforcement
high-yield reinforcement T (and X) bars —0.13%bh
mild steel reinforcement R bars —0.24%bh

Secondary (distribution) reinforcement
high-yield reinforcement T (and X) bars —0.13%bh
mild steel reinforcement R bars —0.24%bh

The clear spacing between bars in slabs should not exceed the lesser of three times the effective depth ($3d$) or 750 mm.

Example

Determine suitable minimum main and distribution reinforcement for a slab of overall depth 270 mm. Use high-yield steel and 25 mm nominal cover (Fig. 19.1(b)).

$b = 1000$ mm
$h = 270$ mm

Main reinforcement
$$A_s = 0.13\% \, bh$$

$$= \frac{0.13}{100} \times 1000 \times 270 = 351 \text{ mm}^2 \text{ per metre width}$$

use T10 @ 200 mm centres (=392 mm²/m)
or T12 @ 300 mm centres (=377 mm²/m).

Distribution reinforcement

As main reinforcement. Although a spacing of up to the lesser of 750 mm or $3d$ ($3 \times 244 = 732$ mm) is permitted, in practice spacings greater than 300 mm are unusual.

Maximum reinforcement

BS8110 does not give requirements for maximum reinforcement in slabs. For the main reinforcement, it is suggested that the requirement for maximum longitudinal steel in a beam is adopted:

$$A_s \not> 4\% bh$$

It is unlikely that there would ever be a need for substantial amounts of distribution steel.

19.3 Beams

Reinforced concrete beams are usually made the minimum depth appropriate to their span and therefore use designed reinforcement, say 16–40 mm bars. However, like slabs, it is occasionally appropriate to make a beam much larger than is necessary for structural needs. In such a case the reinforcement required for load carrying may be negligible.

These notes apply to simple rectangular beams; more complex sections are detailed in the Standard.

Minimum reinforcement

BS8110 specifies minimum areas for the main tension reinforcement and shear links in a beam (Fig. 19.2).

Main (tension) reinforcement
high yield reinforcement T (and X) bars —0.13%*bh*
mild steel reinforcement R bars —0.24%*bh*

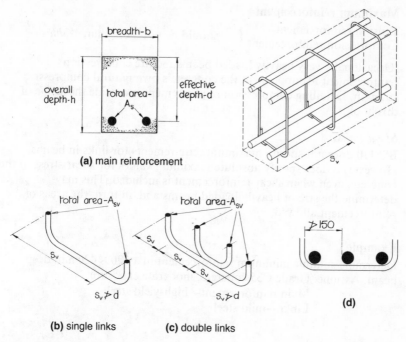

(a) main reinforcement

(b) single links   **(c)** double links

(d)

Fig. 19.2

Shear links (stirrups)

In heavily loaded beams where shear stresses are high, designed links are normally required to supplement the concrete's relatively low shear strength and the Standard details the method of calculating the amount required for particular situations. A link arrangement may be single links (b) or double links (c). In areas of low shear stress, it may be possible to use minimum links for which the area and spacing are based on the size of the concrete profile.

It is often possible to use minimum links along the whole length of moderate and lightly loaded beams.

A_{sv} = total cross-sectional area of links at the neutral axis at a section.

b_v = breadth of section.

d = effective depth

s_v = spacing of links along the beam

$$A_{sv} \geq \frac{0.4\, b_v\, s_v}{0.87\, f_{yv}}$$

The spacing of links along the beam should not exceed $0.75d$. At right angles to the span the spacing of the vertical legs should not exceed d and no tension bar should be more than 150 mm from a link (d).

158

Maximum reinforcement

Tension reinforcement
Compression reinforcement $\Big\}$ should be not more than 4%bh

(Occasionally very heavily loaded beams require compression
reinforcement to supplement the concrete's own natural compressive
strength. Detailing compression reinforcement is outside the scope of
this book.)

Shear links

BS8110 does not specify maximum requirements for links in beams.
However, it does specify absolute maximum values for shear stress in the
concrete, even when shear reinforcement is included. This may
determine the size of heavily loaded beams and, in turn, the areas of
reinforcement allowed.

Example

Determine suitable minimum reinforcement in a 320×420 mm deep
beam. Assume: Grade C35 concrete, moderate exposure.

Main reinforcement—high-yield steel
Links—mild steel

Main reinforcement

$$A_s = 0.24\% \ bh$$

$$= \frac{0.24}{100} \times 320 \times 420 = 323 \text{ mm}^2$$

try 3T12 (=337)

Links

$$A_{sv} \geq \frac{0.4 \ b_v s_v}{0.87 \ f_{yv}}$$

An easy way to use this formula is to assume a link size and then
calculate a suitable spacing.
Rearranging the formula:

$$s_v \vartriangleright \frac{0.87 \ f_{yv} \ A_{sv}}{0.4 \ b_v}$$

Try 10 mm single links:

$\left.\begin{array}{l} A_{sv} = 2 \times 78 = 156 \text{ mm}^2 \\ f_{yv} = 250 \text{ N/mm}^2 \\ b_v = 320 \text{ mm} \end{array}\right\}$ $s_v \vartriangleright \dfrac{0.87 \times 250 \times 156}{0.4 \times 320} = 265 \text{ mm}$

use R10 single links @ 250^c/$_c$

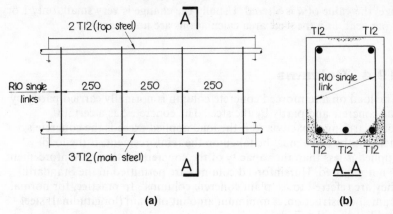

Fig. 19.3

A detail of the proposed reinforcement detail is shown in Fig. 19.3.
  Check spacings:
(i) Along the span $s_v \not> 0.75d$
    Determine d: for a grade C35 concrete with moderate exposure,
    nominal cover = 35 mm

$$d = 420 - 35 - \underset{\substack{\text{link} \\ \text{diam.}}}{10} - \underset{\substack{\text{main} \\ \text{bar} \\ \text{radius}}}{\frac{12}{2}} = 369 \text{ mm}$$

h cover

    Maximum allowable spacing $= 0.75d = 0.75 \times 369 = 277$ mm
    Proposed spacing $s_v = 250$ mm (<277): OK

(ii) Lateral spacing of the vertical legs

$$= \underset{b}{320} - 2(\underset{\substack{\text{cover}}}{35} + \underset{\substack{\text{link} \\ \text{radius}}}{\frac{10}{2}}) = 240 \text{ mm } (<369): \text{OK}$$

(iii) Maximum distance for longitudinal bars is based on the central bar
    = 115 mm (<150): OK.

Top steel

In this case, there are no specific requirements for the top steel, 2 T12s
would be satisfactory.
    In the calculations, the value of the effective depth d is affected by
the size of the proposed bars. If the actual steel used is of a different

size, the value of *d* is altered. Usually the change is very small, only 1 or 2 per cent, and the steel area calculations are not revised.

19.4 Columns

The load on a reinforced concrete column is normally carried partly by the concrete and partly by the steel. The concrete has a certain load-carrying capacity and if the load applied exceeds the concrete's stength, the extra must be taken on the reinforcement. If the load applied is less than the capacity of the concrete, in theory reinforcement is not required. Unreinforced columns are permitted in the Standard: they are referred to as 'plain concrete columns'. In practice, for normal framed construction, a minimum amount of main (longitudinal) steel and links are detailed.

Typical column reinforcement is shown in Fig. 19.4.

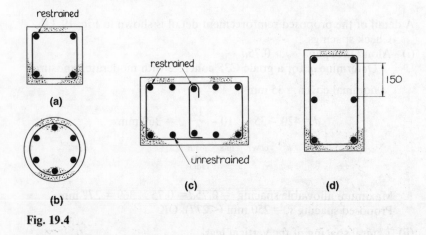

Fig. 19.4

Minimum reinforcement

Main (longitudinal) reinforcement
high-yield reinforcement T (and X) bars
mild steel reinforcement R bars } $A_{sc} = 0.4\%bh$

It is normal practice to detail not less than four bars in a rectangular (or square) column (a) and not less than six bars in a circular column (b). The minimum bar size used is 12 mm.

Links
Size—at least $\frac{1}{4} \times$ size of the largest compression bar.
Maximum spacing—$12 \times$ size of the smallest compression bar.

Every corner bar supported by a link } Fig. 19.4(c)
Every alternate bar supported by a link }

Unrestrained bars must be no more than 150 mm from a restrained bar or group (d).

Links in circular columns normally take the form of helical reinforcement. This is usually supplied to site as a coil and during the fixing is extended to a helix with an appropriate link spacing.

Example

If the main reinforcement in the column (a) is 4 T16s then the minimum links would be chosen as:

Size: at least $\frac{1}{4} \times 16$ mm = 4 mm—propose 6 mm mild steel
Maximum spacing: 12×16 mm = 192 mm—use 175 mm

(In practice, the use of 6 mm bars is not encouraged for *in-situ* concrete work since they are not robust enough to withstand normal construction site conditions. The links in the example would probably be made R8s).

Maximum longitudinal reinforcement

The BS8110 requirements for maximum permitted longitudinal reinforcement are expressed as a percentage of the cross-sectional area of the column:

6% in vertically cast columns
8% in horizontally cast columns
10% at a lap in vertically and horizontally cast columns.

In vertically cast columns, the concrete is normally placed and compacted from the top of the shutter. To give adequate access, the maximum of 6 per cent longitudinal steel is specified. Horizontally cast columns are pre-cast. One side of the shutter is open. Access to place and compact the concrete is better than for vertically cast columns, hence the increased allowance to 8 per cent. The allowance of 10 per cent at laps is permitted because laps cause only local congestion.

Maximum links

The Standard does not specify maximum areas for links in columns. Bar sizes and spacings should be related to the size of the column and the main reinforcement.

19.5 Walls

Concrete walls behave rather like columns. The concrete has a certain load-carrying capacity and if the load applied exceeds the concrete's strength, the excess must be taken on the reinforcement. If the load applied is less than the capacity of the concrete, then reinforcement is not required. Unreinforced walls are permitted in the Standard; they are referred to as 'plain concrete walls'. But like columns, for normal framed construction, a minimum amount of steel is usually detailed.

Minimum reinforcement

Main (longitudinal) reinforcement

high-yield reinforcement T (and X) bars
mild steel reinforcement R bars $\Big\}$ $A_{sc} = 0.4\%bh$

It is normal to design a typical one-metre length of wall (Fig. 19.5(a)) with $b = 1000$ to coordinate with the bar area chart where reinforcement is listed 'per metre width'. The chosen steel is then appropriate to the full length of the wall.

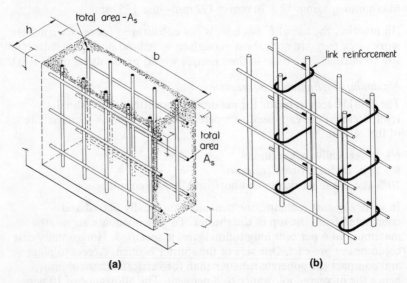

(a) *(b)*

Fig. 19.5

Horizontal reinforcement

$A_s = 0.25\%bh$ for high-yield bars
$A_s = 0.30\%bh$ for mild steel bars.

Size of the bars to be not less than ¼ size of the vertical steel, and not less than 6 mm.

This reinforcement may be shared between two faces.

Maximum reinforcement

Vertical reinforcement

$$A_{sc} \not\triangleright 4\%bh$$

Horizontal reinforcement

No maximum requirement is given in the Standard.

Because walls have large cross-sectional areas, they are not likely to be as highly stressed as columns. Horizontal wall reinforcement is normally 8, 10 or 12 mm at 100 to 300 mm centres, and steel in these ranges is not likely to cause congestion.

Links in walls

In heavily loaded walls, the vertical reinforcement may try to buckle. When more than 2 per cent of the steel is being used to resist compression, a link arrangement must be introduced to give restraint.

Typical wall link reinforcement (Fig. 19.5(b)) uses shape code 85 links which are threaded into the reinforcement cage when the rest of the steel has been fixed. Link spacings and positions are detailed in the Standard.

19.6 Water-retaining structures

Concrete water-retaining structures are designed to BS8007 and because of the contact with water, they are particularly sensitive to the effects of cracking. The Standard includes methods of calculating minimum reinforcement based on the strengths of the concrete and steel. The walls and slabs used for water-retaining structures are often very thick, and to accommodate the different thermal and shrinkage effects which occur, they may be considered to be divided into surface zones and a core. For calculating the minimum reinforcement, only the surface needs to be considered. For walls, these are assumed to be a maximum of 250 mm thick. Walls up to 500 mm thick require reinforcement appropriate to the full thickness. Thicker walls only require reinforcement appropriate to a 500 mm section (Fig. 19.6). Ground slabs may also be divided into a similar way.

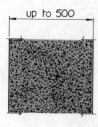

(a) reinforcement based on the full thickness

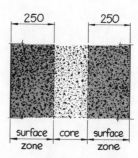

(b) reinforcement based on the two surface zones

Fig. 19.6

Thick walls and slabs in other (non water-retaining) structures have also been reinforced using the surface zone/core division, the minimum reinforcement supplied being based on the percentages appropriate to the particular type of structure.

Section 20

Reinforcement drawings

20.1 Introduction

Reinforcement drawings show the arrangement of reinforcing bars in a concrete structure. The design calculations prepared for a project determine the principal reinforcement in each element, and summaries of the steel required are given at appropriate stages, often with explanatory sketches. The working drawings must show complete reinforcement assemblies, which include the designed reinforcement together with all the additional lacing and trimming bars that cannot be shown in the calculations. Much of the reinforcement required can only be determined as a drawing develops.

20.2 Presentation

The basis of a reinforcement drawing is a simple outline drawing showing the structural concrete profiles, fully detailed to include holes, pockets, cut-outs and so on. The outlines are drawn with a fine line (typically 0.25–0.35 mm) and they are not normally dimensioned, since that information is available on the accompanying layout drawings, though ordnance datum level markers are used. The reinforcement details are then drawn on the concrete profiles. The presentation is diagrammatic, with the bars shown as simple straight and profiled lines, and identified (called up) using standard formats.

Reinforcing bars are cut and bent to the range of standard shapes described in section 15. The bent size of a bar is calculated from the concrete profiles, with suitable allowances made for concrete cover, laps between bars and so on. Bars drawn in elevation are shown as single lines, using the same line thickness regardless of the actual bar diameter. The lines are normally 0.35–0.5 mm, depending on the scale of the drawing. In cross-section, bars are shown as single dots or filled circles, the choice also depending on the scale of the drawing. The circles can be drawn with a circle template. The presentation of some typical bar shapes is shown in Fig. 20.1.

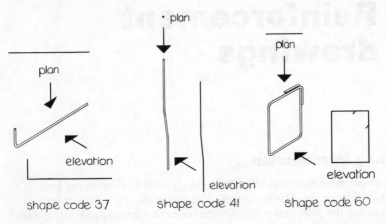

Fig. 20.1

Even a simple reinforced concrete structure may contain several hundred separate bars but it is not necessary to show them all on the drawings. Reinforcement is normally detailed as bars in a group, or as bars in a string placed in parallel at defined spacings. Standard methods of layout and presentation are used. Bars detailed in a group are normally drawn with one in full profile together with a short dash to show the extent of the group (Fig. 20.2(a)). An identifying line links the bars and is drawn clear of the concrete outline where the reinforcement can be called up.

For the column main steel shown in the figure:

$$4 \quad T20 - 01$$

number of bars in the group ⟋   ↑   ⟍ bar mark number

bar type (mild or high-yield steel) and size

The exact location of each bar in the group is shown on an accompanying cross-section.

Bars detailed in a string are shown in a similar way but the call-up includes the bar spacing (b). Strings may also be shown as two dashes to

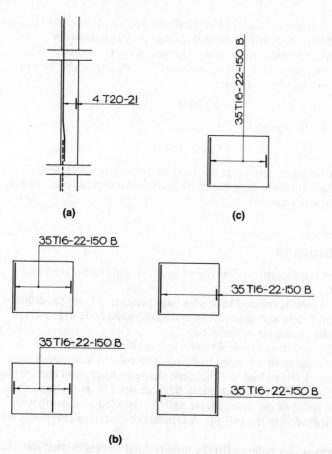

(a)

(b)

(c)

Fig. 20.2

define the extent, and an intermediate full profile bar. For the slab reinforcement shown in the figure:

35 T16 – 02 – 150 B

number of bars in the string

bar type (mild or high-yield steel) and size

bar mark number

location (in the slab)

spacing of the bars in the string

The location may be further identified using an alpha/numeric form B1, B2, etc. to indicate the relative position of bars in mats of reinforcement.

Using arrow heads and dots on the identification line is common practice but other techniques are also used. The way these lines are taken clear of the concrete profile will depend on the particular details, and the layouts shown in (b) are all equally acceptable. The bar identification is written horizontally, which is preferred since the presentation is easier to read, but vertical call-up can also be used (c). As details are added, the reinforcement drawings may become very congested and so a clear, open style of presentation is essential.

Alternative methods of call-up are used. For example, the bars in the column may be identified as:

4.T2001

and the bars in the slab as:

35. T1602. 150 B

The reinforcement drawings are read in conjunction with bar schedules which tabulate each bar mark, giving full dimensional details and the quantities required.

20.3 Columns

Basic column reinforcement consists of longitudinal (main) steel and links.

The main steel is secured to starter bars projecting from the column or foundation below and is normally cranked (shape code 41) to keep the bars on the perimeter of the cage.

Figure 20.3(a) shows typical reinforcement details for an intermediate column in a framed building. The column contains four main bars (mark 01s) which are the same shape and size, and rectangular links (shape code 60). The links (mark 02s) are used from the top of the kicker to the soffit of the beam above and are detailed as a string with the spacing included in the call-up. A typical cross-section completes this detail.

In the top-storey column (b) the longitudinal steel is carried into the beam. The joint is made with separate L-bars which allows maximum flexibility when the steel is fixed. An alternative main bar (c) can be used but this is cumbersome and may be difficult to fix to a specific level needed to achieve the top cover. The bar may also foul on the interconnecting beam steel. The column (d) is cast with pockets to take dowel fixings and a pre-cast concrete beam (inset). Although the concrete stresses in the head of the column may not be high, additional links have been included around the pockets to generally strengthen the area.

These examples show simple rectangular columns with four main bars and strings of single links, but columns often contain six, eight or more bars and may be a complex shape requiring multiple link arrangements. In each case the column cage must be stable with the longitudinal bars adequately braced to the requirements of the design codes.

Figure 20.4 shows a number of column sections with longitudinal and link reinforcement.

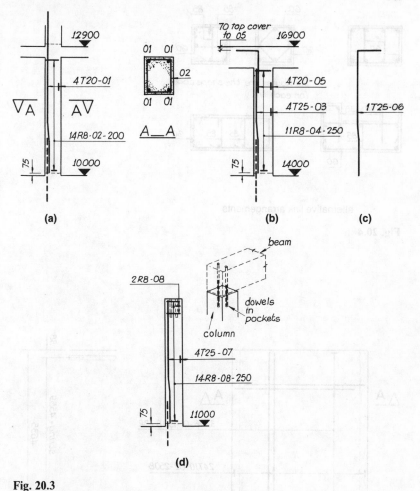

Fig. 20.3

When the main bars are of different sizes, they should be set out symmetrically. When multiple link arrangements are used, the vertical spacings should be the same, allowing links in a group to be wired together. Wherever possible, one link should be used to form the external shape of the cage with the others added to give the required bracing.

20.4 Slabs

The single-span slab shown in Fig. 20.5(a) is reinforced with designed main steel in the direction of the span (mark 01) and minimum distribution steel at right angles (mark 02). The main steel is designed to carry the bending moments caused by loading effects. The size of the

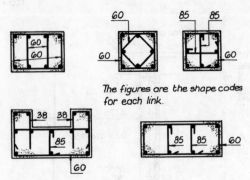

The figures are the shape codes
for each link.

alternative link arrangements

Fig. 20.4

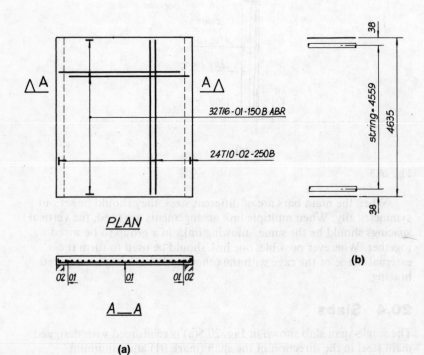

PLAN

32 T16 - 01 - 150 B ABR

24 T10 - 02 - 250 B

string - 4559

4635

38

38

A__A

(a)

(b)

Fig. 20.5

moments varies along the span and the variations are shown in structural calculations as bending moment diagrams.

In this simple case, the reinforcement design is based on the maximum moment which occurs at or near the middle of the span. The steel is detailed as strings of parallel bars and the identification includes the centres (150/250 mm) and the location (B = bottom). As the bending moments reduce towards the supports it is economic to reduce also the amount of reinforcement detailed. Only 50 per cent of the maximum reinforcement needs to be carried onto the support and suitably anchored, the remainder may be stopped short (curtailed). On the reinforcement details this is achieved by using one barmark for all the main steel. The bar is detailed to be shorter than the span and has a bob at one end (shape code 37). Placing the string of bars with the bobs alternately over the left and right hand supports fulfils both the anchorage and curtailment details.

The bar call up includes the initials ABR (alternate bars reversed). On the typical section, two reversed 01s are shown and the ends indicated with angled tails. This is a detailing technique to show the extent of the bars. They are not actually bent with the ends angled. The number of bars required in the string is calculated from the chosen bar spacing which must be a practical dimension. The overall width of the string is taken as the slab width, less edge distances, to the first and last bars. It is unlikely that the chosen spacing will divide exactly into the overall width and the number of bars should be rounded up to accommodate this. For example, if the slab is 4635 wide and is reinforced with T16 bars at 150 centres, with 30 cover:
Referring to Fig. 20.5(b).

Overall width of the string $= 4635 - 2\,(30 + \dfrac{16}{2})$

$$= 4559$$

Number of spacings $= \dfrac{4559}{150} = 30$ remainder 59.

On the drawings, detail thirty + one for the remainder + one to start the string = 32.

$$32T16 - 01 - 150 \; (B. \; ABR)$$

Exactly spaced, these bars would be at $\dfrac{4559}{31} = 147$ centres which would be impractical to fix. On site, the steel fixer can place most of the bars at 150 centres by closing the spacings of a few at each end of the string. The number and spacing of the distribution bars is determined in the same way.

Holes are often required in slabs to accommodate pipe-runs, air-conditioning ducts and other services (Fig. 20.6). Small holes (less

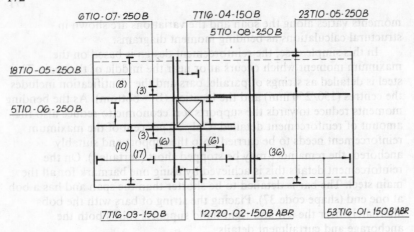

Fig. 20.6

than about 150 mm square) can normally be formed in the slab without modification to the basic reinforcement details, but for larger holes special provisions may be required. The structural design is then based on the assumption that the areas of slab on each side of the hole span onto 'beam strips' formed within the adjacent slab (Fig. 20.7(a)). The beam strips are strengthened to carry the effects of these redistributions of the slab loading. It is often sufficient to replace the area of steel cut by the hole with an equivalent area shared between two beam strips (b). In the example shown, the T16-01s have been replaced by T20-02s over an appropriate width. Additional distribution reinforcement— T10-07s—has been detailed to compensate for the T10-05s cut by the hole. This detail is suitable for holes of side length up to about 0.5 m. For larger holes (side length up to about 1 m) the same basic detail may be used but bars of an area equal to the cut steel would also be used in

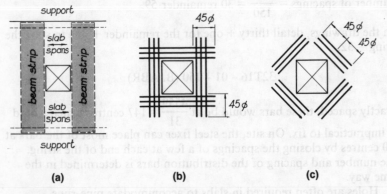

Fig. 20.7

each direction in the top of the slab. The 45ϕ is the nominal anchorage length. In addition, when the slab depth exceeds 250 mm, bars of a similar area should be placed diagonally across the corners of the hole (c).

For very large holes, the beam-strip method may be inadequate and special provisions such as framing beams may be required. Suitable designs are then considered for each situation.

When a slab is continuous over several supports loading causes sagging moments in the spans and hogging moments over the supports. The critical designed reinforcement is detailed in the bottom of the spans and in the top over the supports. The steel can be reduced or curtailed away from the areas of critical bending moment. The stopping-off points can be determined by calculation. Alternatively, simplified rules are given in BS8110 which cover curtailment of steel in slabs of approximately equal spans which fulfil certain basic loading conditions. Cantilever slabs are reinforced to the bending moment diagram with critical top steel at the cantilever support and bottom steel in the adjacent span. Simplified rules for curtailment are also given in BS8110.

In most structures the effects of shear are not a problem when the slab is carried on continuous supports such as walls or beams.

20.5 Beams

Beams are normally reinforced with longitudinal bottom and top steel and links. The simple, single-span beam shown in Fig. 20.8 is reinforced to carry a light, uniformly distributed load and the steel is designed according to the bending moment and shear force diagrams. The detail drawing normally shows an elevation and a section. The main (bottom) steel (mark 01) is designed to the maximum moment which occurs at mid-span. The bars are continuous to the supports where they are

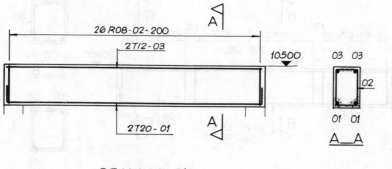

BEAM Mk 5/1

Fig. 20.8

anchored to the requirements of the design codes, usually by taking them twelve diameters beyond the centre of support. When the shear stresses in the beam are low and can be carried by the concrete alone, nominal shear links (mark 02) are provided. The links are detailed as a string and the number of bars required is determined in the same way as strings of bars in slabs. The cage is completed with longitudinal top steel (mark 03) which, in this case, is not designed. Instead, bars are used which are one or two sizes smaller than the main steel.

For a more heavily loaded beam (Fig. 20.9) the reinforcement details are adjusted to accommodate variations in bending moments and shearing forces. The main steel (mark 04, 05) is designed to the bending moments, starting with the critical maximum value, which is often at or near mid-span. As the moment reduces towards the supports, the area of steel is adjusted either by stopping some of the bars, (mark 05) or by changing to bars of a smaller diameter, which involves a lap. The cut-off points for the bars are critical, and fully detailed sketches should be included in the structural calculations. The links are designed from the shear force diagram. Nominal links (mark 06) are detailed for those parts of the beam where the shear force is low and the concrete can carry the resulting stresses unaided. Areas of higher shear require designed reinforcement which may be detailed either using the same (nominal) links but closing the centres to increase their effectiveness, or by using links of a larger diameter at appropriate centres (mark 07). Full details should be given in the structural calculations. The reinforcement cage is completed with top steel as before. The reinforcement detail drawing normally shows an elevation and several sections chosen to cover each different steel arrangement. Very heavily loaded beams may require substantial reinforcement (Fig. 20.10).

The main steel (mark 09, 11) may be placed in two or more layers, with short lengths of bar used as vertical spaces (mark 10). The

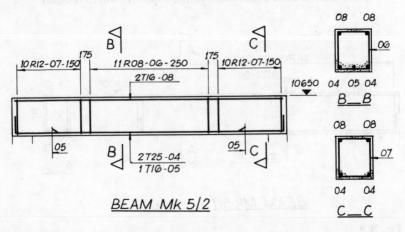

BEAM Mk 5/2

Fig. 20.9

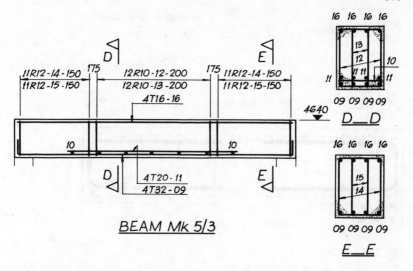

Fig. 20.10

horizontal and vertical spacing of the main steel is critical since
congested reinforcement can lead to poor concrete compaction. Multiple
links are detailed to cater for the very high shear stresses which occur in
this type of beam. One link in a group should be used to form the
perimeter of the cage (mark 12, 14). Often in heavily loaded beams the
flexural compression load exceeds the strength of the concrete. The top
steel is then designed as compression reinforcement to supplement the
concrete's strength.

Beams continuous over several supports are detailed according to
the bending moment and shear force diagrams with principal
longitudinal reinforcement placed in the bottom between spans and in
the top over the supports. Cantilever beams are also reinforced to the
bending moment and shear force diagrams. The reduced areas of steel
sufficient in areas of low stress can be determined either by calculation
or in many cases by using simplified rules for curtailment detailed in
BS8110. These are similar to the rules used for slabs. Reinforcement
details for these beams are normally shown as an elevation and sections.

Beams are often cast with columns or other beams to form
continuous frameworks, or they may be cast with a slab over to form a
beam and slab floor construction. Connections in these structures
require careful detailing since many bars are brought together in a
confined space. They are often of large diameter and this causes
problems of alignment. In the three-beam structure shown in Fig. 20.11
each beam is detailed with longitudinal top and bottom steel and links.
The longitudinal bars in the secondary beam are stopped short of the
connection and the link with the primary beams is made with U-bars.

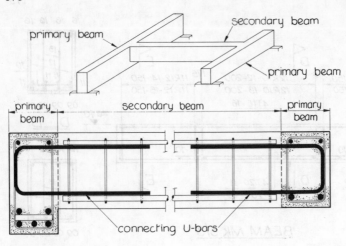

Fig. 20.11

This simplifies the steel fixing since each beam cage is independent and can be pre-fabricated prior to assembly in the shutter. The connecting U-bars fit neatly inside the beam cages.

20.6 Stair flights

The simple stair flight (Fig. 20.12) is designed as a slab spanning between the centres of the supporting walls, with the main reinforcement in the bottom, designed to accommodate the moments on the bending moment diagram. The effect of shear force are unlikely to be a problem for simple stairs.

Reinforcement details are shown in plan and section (Fig. 20.13). At the bottom of the flight, the main steel (mark 01) continues onto the landing. At the top it is not acceptable to bend the bars around an inside corner and so the joint is made by crossing the flight and upper landing

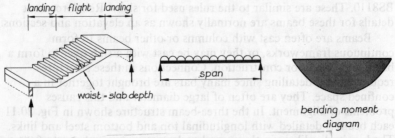

Fig. 20.12   The South East Essex
College of Arts & Technology

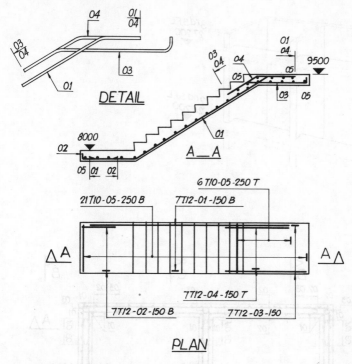

Fig. 20.13

reinforcement (mark 03), locking off each bar with an appropriate anchorage length. A cranked top bar (mark 04) is also included to tidy the joint detail although this is not essential to the flexural strength of the slab. An enlarged detail showing the layout of bars in the top joint is shown. The reinforcement details are completed with distribution steel (mark 05), based on the minimum slab requirements. Although a full plan has been shown for completeness, simple stair flights can be fully detailed on the section alone.

20.7 Walls

Solid cast *in situ* concrete walls are often included in building frames. They are used for gables and around lift shafts and give lateral stiffness to a frame, enabling it to resist wind loads and other effects.

These walls are normally reinforced with mats of vertical and horizontal steel. The details shown in Figs 20.14 and 20.15 are for a lift shaft. Each section of the wall is drawn in elevation to show the layout of the reinforcement and a plan is drawn to show the location of the bars within the wall section. A normal storey-height wall is cast in a

178

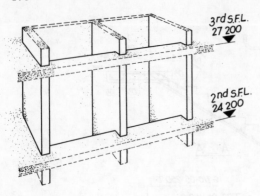

Fig. 20.14

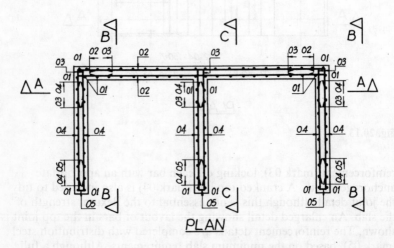

PLAN

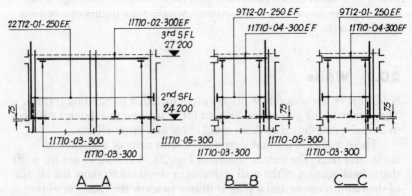

Fig. 20.15

single lift. The vertical reinforcement is spliced onto bars projecting from the wall below and is made long enough to form starters for the continuation of the wall above. The corners are formed with U-bars (mark 03—shape code 38) which form a neat and simple detail (Fig. 20.16(a)). U-bars are also used to close the ends of each wall (mark 05). Alternative corner details (b) and (c) can also be used. In each case the vertical spacing of all bars forming the joint should be the same. If the legs of the L-bar are of similar length, then the bar could be accidentally fixed the wrong way around in the cage. This possibility is overcome by detailing both legs to be the same (longer) length.

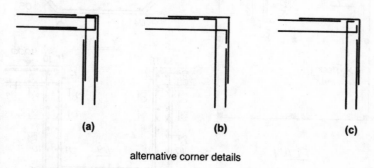

(a) (b) (c)

alternative corner details

Fig. 20.16

The basic mats of horizontal and vertical steel are detailed as strings and the number of bars required in each is determined from the concrete profiles, using the methods used for slab steel. For the type of wall shown, the vertical steel is normally light or moderate, typically T10/12/16 bars at 150–300 mm centres and the bars are not cranked at the lap since there is adequate space to fix them beside the starters.

20.8 Ground slabs

Ground slabs are normally cast directly onto a compacted hardcore sub-base which has been blinded and covered with heavy gauge polythene sheet. The slab is not required to span and flexural reinforcement is unnecessary, but sheets of fabric are placed in the top to minimise the effects of shrinkage and cracking. Ground slabs normally incorporate a number of special features such as edge details for door thresholds, springings for stair flights and the ducts and pits required for services. These are normally reinforced with simple cages of light steel (T10s, T12s, etc.) which can be easily laced to the basic fabric reinforcement.

Figure 20.17 shows typical reinforcement details for a ground slab in a simple rectangular building, including a door threshold (section X–X)

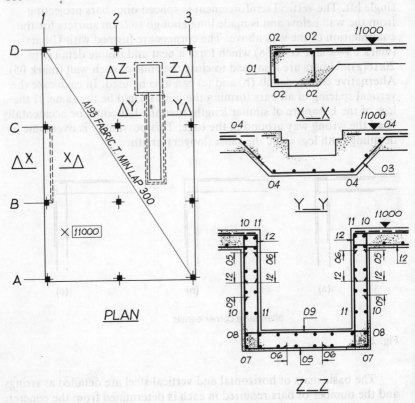

Fig. 20.17

a service duct (section Y–Y) and a pit (section Z–Z). On the details, the fabric is shown by a single line drawn diagonally across the slab. The direction of the line indicates the location of the fabric—top or bottom (Fig. 20.18(a), (b)).

20.8 Ground slabs

Ground slabs are normally cast directly onto a compacted hardcore sub-base which is then blinded and covered with heavy gauge polythene sheeting. Steel is not required to span and flexural reinforcement is unnecessary, but sheets of fabric are placed in the top to minimise the effects of shrinkage and cracking. Ground slabs normally incorporate a number of special features such as edge details for door thresholds, upstands for stair treads and the ducts required for services. These are normally provided with small pieces of light steel (T10s, T12s, etc.) which can be easily fixed to the basic fabric reinforcement.

Figure 20.17 shows typical reinforcement details for a ground slab in a simple rectangular building, including a door threshold (section X–X),

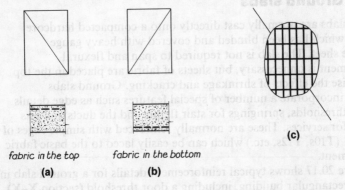

fabric in the top   fabric in the bottom

(a)   **(b)**

Fig. 20.18

The call-up includes the fabric type (A192, etc.) and the minimum lap between adjacent sheets. Normally an A-series square mesh fabric is used and there is no need to specify the orientation. When rectangular meshes are detailed, the orientation can be shown as a small inset sketch on the plan (c). The typical sections are used to show the exact extent of the fabric at slab edges, ducts, etc. The separate bars (mark 01–12) should also be fully detailed on the plan but are omitted in this example for clarity. The requirements for cover to the reinforcement, including the mesh, would be detailed in the notes box on the ground slab drawing.

20.9 Pad foundations

The simple pad foundation (Fig. 20.19(a)) works like an inverted cantilever slab with the principal reinforcement in the bottom, tensile face. Provision must also be made for the column that is cast on the pad and a cage of starter bars and links is detailed.

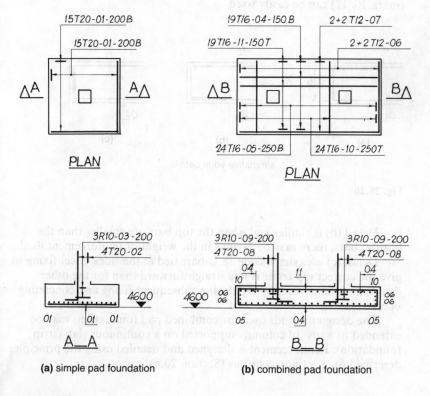

(a) simple pad foundation   (b) combined pad foundation

Fig. 20.19 (a)

Normally the reinforcement forming the bottom mat is made the same in both directions (mark 01). Straight bars may be used, but often the steel must be bobbed to give the additional length required for a tension anchorage to develop. The starters (mark 02), one for each bar in the column, are detailed to project above the top of the kicker by the required lap length and to give stability during fixing, the horizontal leg is made long enough to span at least two bars in the base mat. The cage is completed with links (mark 03).

The combined pad foundation (b) supports two columns and works as an inverted slab with critical tensile zones in the bottom beneath the columns and in the top between the columns. The foundation is reinforced with a full cage of top and bottom steel. Lacing bars are fixed around the sides (mark 06/07). Column starters and links (mark 08/09) complete the reinforcement details.

Alternative edge details are shown in Fig. 20.20. The detail (a) shows the arrangement used for the combined pad. The bottom bars (mark 04/05) are bent to shape code 55 with the B and D dimensions calculated to give the correct top and bottom cover in the completed cage. The bottom-steel mat forms a rigid base on which the top steel (mark 10/11) can be easily fixed.

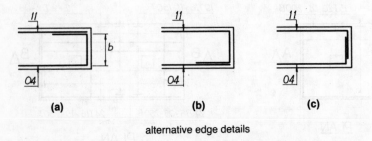

alternative edge details

Fig. 20.20

Detail (b) is similar and when the top bars are smaller than the bottom bars, there may be savings in the weight of reinforcement used.

Detail (c) uses shape code 38 U-bars tied at the sides. Steel-fixing to give the correct top cover is less straightforward than for the other details and the joints may slip during subsequent fixing and concreting operations.

The design methods used for combined pad foundations may be extended to a row of columns supported on a continuous slab (strip foundation). Reinforcement is designed and detailed using the principles described for continuous beams (Section 20.5).

Part 3

Section 21

Structural steelwork

21.1 Introduction

Steel, in the form of standard rolled and drawn sections, is commonly used in both building and civil engineering for the construction of complete frames. Used in this way it is referred to as *structural steelwork*. It is normally made by specialist contractors (steel-fabricators) who make the individual elements of the frames—beams, columns bracings, etc.—in their workshops. These are then taken to the site where they can be quickly assembled into a finished structure. Other building operations such as the construction of floors, deckings, claddings, services and finishings are often carried out as sections of the frames are completed. Long-span steel structures such as roofs and bridge decks are formed without the need for the expensive temporary support systems often required with other forms of construction.

This section describes the materials, construction and methods of design and detailing used for structural steelwork.

21.2 Steelwork in structures

Steel building structures are normally assembled from sections produced to a range of standard profiles. The sections are hot-rolled from steel ingots and vary in size and shape from 25 × 25 mm angle to I-sections almost a metre deep. The profiles are listed in British Standards.

Fig. 21.1 (a)

(b)

Figure 21.1(a) shows a large stanchion-and-truss building erected on a green-field site. The building is single storey and consists of welded steel roof trusses supported on stanchions. There is also an extensive system of secondary bracing. Buildings of this type are normally used for factories and warehousing and are clad with proprietary profiled metal sheeting on both roof and walls.

The beam-and-column building frame (b) is built on a redevelopment site in an urban area. The floors are constructed using a proprietary system of pre-cast concrete units placed dry and then screeded to produce a surface suitable for carpets or thermoplastic tiles. The building will eventually be clad in brickwork, and much of that weight is also carried by the frame. The heavy loadings mean that substantial beam and column sections are required. In building frames of this type, the suspended concrete floors may also be formed using proprietary profiled metal decking. Laid between the steel floor beams, the decking acts as permanent shuttering for the cast *in situ* concrete. It then becomes the principal reinforcement in the completed floor slabs. A light fabric is used as supplementary reinforcement. Special studs (shear connectors) are welded through the decking to the top flange of the beam before the concrete is cast, which makes a full structural connection between the beam and slab. This arrangement, which is structurally very efficient, is called *composite construction*. Ease and speed of erection is an important aspect of this building technique.

A range of cold-formed steel sections is also used with building frames. They are made by drawing flat steel strip through dies to produce the required section which is normally a Z-profile light enough to be carried by hand. Used as purlins, they are the primary support system for profiled metal claddings of the type used on the warehouse building.

The steelwork used in civil engineering structures such as bridges and industrial works may also be constructed from the range of standard steel sections.

These structures are often heavily loaded, and, if they carry cranes and other machinery, there may also be dynamic loading. This may require rolled sections from the heavy end of the range, and occasionally, compound sections made by welding or bolting standard sections together.

Often, however, sections are required which are larger than the standard range, and these are produced by welding flat steel plates together to form welded plate girders (beams) and columns. The bridge deck beams shown in Fig. 21.2(a) are fabricated in this way. Shear connectors are welded to the top flanges. The structure is completed by casting on a concrete slab which then works structurally with the beams. Shuttering and reinforcement are already in place for part of the slab. This is another example of composite construction.

The welded plate girders (b) have been delivered to site by lorry and are being erected as part of a composite construction bridge deck.

Fig. 21.2 (a)

(b)

21.3 Steel quality

The qualities of steel used for construction are defined in BS4360: 1986: Weldable Structural Steels. They are classified by tensile strength and chemical composition.

The tensile strength of a steel may be determined by loading a standard specimen of the material in a tensile test machine. Load is applied to the specimen in increasing increments and the resulting extensions are measured very accurately. A typical load/extension curve for hot-rolled steel is shown in Fig. 21.3. It is similar to the curve obtained for the hot-rolled steels used in reinforced concrete.

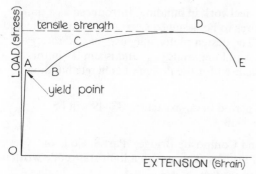

Fig. 21.3

Tensile failure occurs in the specimen at the point of maximum load (D) and the stress appropriate to this load is used to classify the grade of steel. The grade is taken as (approximately) one-tenth of the minimum tensile strength in N/mm^2. For example:

Grade 43—minimum tensile strength 430 N/mm^2. This is a 'mild steel' and is commonly used in structural steelwork fabrications.
Grade 50—minimum tensile strength 490 N/mm^2. This is a high-yield steel used for bridgeworks and increasingly in building structures.

Each grade is subdivided and further identified by a letter A–F. The letters indicate a range of chemical composition and strength characteristics. The different steels are suitable for particular applications. Some are used for the very thick plates used in heavy civil engineering; others are suitable for low-temperature environments or offshore applications. Although BS4360 defines an extensive range of steels, the most commonly used in the construction industry are grades 43A and 50A. The standard also describes a range of 'weathering' steels which can be used for structural work in exposed situations without an applied corrosion protection system. The chemical composition allows limited initial corrosion to take place. An oxide film forms on the surface which then protects the steel from further corrosion.

Weathering steels originated in the USA. In Britain they are classified as grade WR50 steels (WR—'weather resistant') but are often known by the American name of Corten.

21.4 Design codes

Structural steelwork may be designed to:

BS449: Specification for the use of structural steel in building. This is an elastic stress code that has been used for many years for the design of steel structures;

BS5950: Structural use of steelwork in building. Introduced as a limit state code allowing structures to be designed separately for ultimate and serviceablility conditions. The design philosophy uses partial safety factors for the loadings (γ_f) and materials (γ_m) and is similar in this respect to BS8110, the code used for the design of concrete building structures.

It is intended that, after a period of co-existence, BS449 will be withdrawn in favour of BS5950.

BS5400: Steel, Concrete and Composite Bridges. Part 3. Code of practice for design of steel bridges. This is also a limit state code written specifically for the design of bridges.

Although the grading of steel is related to the tensile stress at failure, this is not the stress which is readily recognised by the design team. Steelwork design to BS5950 uses the design strength (p_y) of the material. This may normally be taken to be equal to the yield strength. Values of p_y for grade 43 and 50 steels are shown in Fig. 21.4, which is taken from Table 6 of BS5950: Part 1: 1985.

Table 6. Design strengths, p_y, for steel to BS 4360		
BS 4360 Grade	Thickness, less than or equal to	Sections, plates and hollow sections p_y
43 A, B and C	mm 16 40 100	N/mm² 275 265 245
50 B and C	16 63 100	355 340 325
55 C	16 25 40	450 430 415

Fig. 21.4

21.5 Steelwork drawings

Responsibility for the design and detailing of the structural steelwork for a project is normally divided between the structural engineer's design team and the steel-fabricator who makes and erects the steelwork.

On building contracts it is normal for the design team to undertake the overall structural design and for the steel-fabricator to design the connections between individual members. At the scheme design stage, the design team carries out a preliminary appraisal of the whole structure, which includes calculating the sizes of principal members in the steel frame and the preparation of scheme drawings. The structural engineer may prepare these drawings to get a 'feel' for the structure. Alternatively, a structural detailer may work with the structural engineer to produce this preliminary information.

Later the design team complete detailed structural calculations, and using the architect's drawings and structural scheme drawings, produce a set of working drawings for the structure. These normally include separate details for the steel framework.

A typical general arrangement drawing for a portal-framed building is shown in Fig. 21.5. The drawing shows the completed frame and details which relate to structural and architectural features. The overall structural dimensions are given, but detailed dimensions of the individual elements and details of the connections between them are not shown. However, the drawing does include (small-scale) line drawings of the frame on which the forces and bending moments at each connection are given.

The steelwork drawings prepared by the design team are often drawn over a period of weeks or months, being worked on and modified as the design develops. They are normally drawn in ink, in a style which matches the other structural drawings. Their long drawing-board life means that they must also be durable—they are usually drawn on good-quality tracing paper or plastic film.

The main contractor normally sub-lets the fabrication and erection of the steel frame. Several fabricators may be invited to submit tenders for the work. The connections between elements in the frame constitute a major part of steelwork costs. The practice by which the fabricator is responsible for the connections allows each one to propose economic joints which accord with their own fabrication and site-erection methods and this constitutes part of the competitive element in the fabricator's tender for the work.

Alternatively, the steel-fabricator may be nominated by the design team. Once the sub-contract has been let, copies of the design team's drawings are sent to the fabricator who needs them to produce 'shop' and general arrangement drawings.

A typical fabricator's shop drawing for elements required in the portal frame is shown in section 30. The drawings show the craftsmen and craftswomen in the workshop exactly what has to be made. This

192

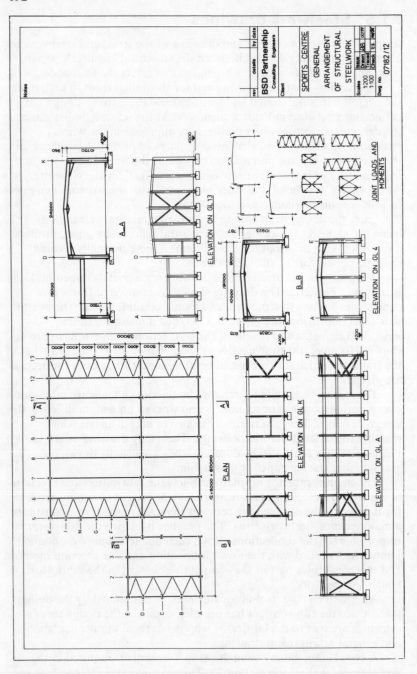

Fig. 21.5

includes details such as the required cut lengths of the basic sections, dimensions of welded-on plates, holes and corrosion protection systems.

Separate drawings are normally prepared for each different group of elements—beams, columns, etc. The fabricator also prepares general arrangement drawings which show the assembled frame and include plan grid dimensions, diagonal dimensions across the grid (used on site to check squareness) and levels (OD) of the erected frame. Many drawings are required, even for a simple building frame.

On receipt of an order the fabricator may require shop drawings very quickly. They are frequently drawn in pencil on tracing paper with the number and letterwork done freehand (not stencilled). Drawing in pencil is quick and modifications are more easily made than to an ink drawing. Fabricators' drawings often have a distinctive italic style, which is not to comply with any code or standard but the result of custom and practice within the industry.

Computerised drafting, detailing and preparation of input data for computer-controlled automatic cutting and drilling machines are also becoming part of the detailer's work.

For civil engineering contracts, a similar design and fabrication procedure is followed but architects are not involved and the project design is supervised by the engineer whose team carry out the preliminary studies and detailed design and prepare the contract documents. It is also likely that the design team will design the connections between members.

21.6 Weight

The steel in a structure is normally quantified by its weight in tonnes, and the weight per square metre of floor (or deck) is an indicator of the economy of a frame.

In building and civil engineering bills of quantities, the supply and fabrication items for structural steelwork are itemised in tonnes.

The cost of different forms of construction is similar. Variations may be due to the complexity of connections and the type of sections used. Lightweight trusses and other fabrications made from small sections such as angles and tubes are normally slightly more expensive than steel frames made using more substantial sections.

Typical steelwork weights are:

	Total weight (tonnes)	Weight in kg/m^2 of floor (or deck)
For the large stanchion-and-truss warehouse (Fig. 21.1(a))	600	25
For the three-storey office building (Fig. 21.1(b))	120	50
For the 40-m span single-carriageway river bridge (Fig. 30.5, page 282)	125	200

21.7 Corrosion protection

Steelwork must be protected from corroding. This may be done by painting, galvanising, spraying with a protective metal coating or by a combination of these finishes. There are also available specially alloyed steels in which limited surface corrosion takes place. The resulting rusted finish inhibits further corrosion. These are the weathering steels described previously.

Structural steelwork may also require fire protection and this may be achieved in many ways, including special paints, spray-on materials, cladding in concrete and enclosure within architectural finishes. The chosen protection system must have the appropriate fire rating.

21.8 Trial assembly

Steelwork is detailed and constructed to very high standards and serious errors are uncommon. When a straightforward building frame is assembled on site, erection problems are normally confined to minor misalignments between the elements. If more serious errors occur, the necessary corrections can often be made *in situ*, although this may cause delays to the contract. For some types of work, delays in the site erection of the steelwork could have serious consequences. For example, installing beams over a railway line requires a track possession, which only lasts a few hours. Similarly, construction over an existing road requires a road closure, which is also timed.

The fabricator can reduce the possibility of delays on site by carrying out a trial assembly of the steelwork in the workshop. This ensures that the elements fit together and allows critical dimensions to be checked. The fabricator may also use the trial erection to build sub-assemblies *in situ*, and to complete through-drilling for bolted splice-joints where accurate drilling of separate components is difficult.

Trial erections are normally compulsory for road and rail bridges and an item is included in the bill of quantities for the contract.

21.9 Transport

Structural steelwork is normally transported from the fabricator's works to the site by lorry.

The size and weight of the loads carried on goods vehicles is controlled by the Motor Vehicle (Construction and Use) Regulations and other legislation, and the restrictions which they impose may influence the way the steelwork is detailed (Fig. 21.6).

No special control is required for goods carried within the length and width of a C+U vehicle, but if the load overhangs, the regulations

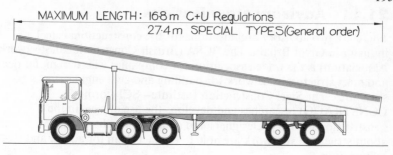

Fig. 21.6

may require warning markers, notice to the police and other controls. Generally, the larger the load, the greater the control. The maximum payload that may be carried is fixed by the permitted gross weight of the loaded vehicle. For certain lorries such as the tractor and the trailer shown in the figure, a maximum gross weight of 38 tonnes applies.

Much larger and heavier loads can also be transported but special vehicles and further controls are required.

The structural elements used in building frames are normally small enough to be transported by road without difficulty, but roof trusses and the beams and girders used in civil engineering works may be too long for transportation in one piece, and so they must be detailed for fabrication in shorter sections which can be site-assembled by bolting or welding.

There are similar restrictions on the size and weight of loads transported by rail.

21.10 Site erection

On site, steel frames are usually assembled with bolted connections.

A lorry-mounted crane is used to lift the steel stanchions, beams and trusses into position and the connections are made by a team of steel-erectors. For bolted connections, they are able to work from ladders. The nuts, bolts and washers are easily assembled by hand and then tightened with a simple hand spanner.

Site connections can be made by welding but this is normally avoided. When it is necessary, metal arc-welding is commonly used. The equipment is bulky and access platforms are needed for working at heights. Welding is affected by the weather, and a temporary enclosure may be required around the working area. Quality control and inspections are also difficult. However, for certain types of work, site welding is necessary and the appropriate controls must be applied.

21.11 Advisory organisations

There are two advisory organisations for the constructional steel industry in Great Britain. The BCSA (British Constructional Steelwork Association) is the trade association for companies who design, fabricate and erect structural steelwork for building and civil engineering structures. The Steel Construction Institute—SCI (formerly CONSTRADO) researches and develops the use of steel in the construction industry. Designers and detailers who have queries about structural steelwork are able to contact them for advice and information.

The British Steel Corporation (BSC) also offers a technical advisory service through its regional offices, which covers all aspects of the design and construction of structural steelwork.

Section 22

Structural steel sections

22.1 Introduction

Steel structures are assembled from a wide range of standard and purpose-made sections.

Rolled sections are made to profiles specified in British Standards and they vary from 25×25 mm L-sections (angles) to I-sections (universal beams), almost a metre deep. The majority of structural frames are made from rolled sections (Fig. 22.1(a)).

Beyond this, it is possible to produce a range of much larger sections. These are made by continuous welding steel plate up to 100 mm thick.

Hollow sections are square, circular or rectangular tubes, which are often used in long-span lightweight roof construction.

Steel purlins and sheeting rails formed from mild steel strip are used as framing to carry lightweight profiled sheet claddings on factories, warehouses and similar types of building (Fig. 22.1(b)).

22.2 Rolled sections

Rolled sections are produced from cast-steel ingots which are initially rolled into rough, square-sectioned bars, called blooms and billets. These are then reheated and passed through profiled rollers which successively reduce the bars to the final structural shapes. Some

Fig. 22.1 (a)

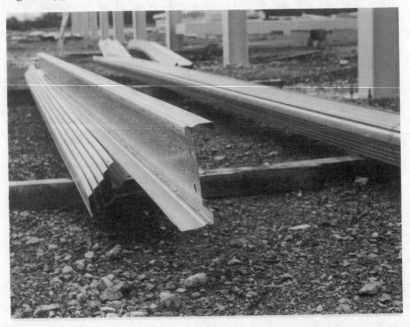

(b)

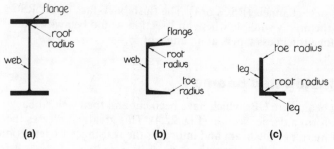

Fig. 22.2

standard terms commonly used with rolled sections are shown in Fig. 22.2. Each standard shape is suitable for a particular application. For steelwork detailing, the shapes are referred to by initials or profiles.

22.3 Typical rolled sections

Cross-sections through typical rolled sections are shown in Fig. 22.3.

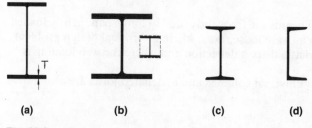

Fig. 22.3

(a) Universal beams (UBs) are used as main-beam members in frames. They may also be used as columns in lightly loaded structures such as single-storey factories and warehouses. The flange thickness 'T' is constant. This is important for detailing.

(b) Universal columns (UCs) are approximately square (see inset sketch) and are used to carry vertical loads because of their resistance to buckling. A (larger) variation on the UC is the steel 'H' pile used in foundation construction. These are called universal bearing piles (UBPs). For UCs the flange thickness 'T' is constant.

(c) Rolled-steel joists (RSJs) or joists are a range of small I-sections. The largest RSJs are about the same size as the smallest UBs. Joists are used for short-span beam and lintel applications and as heavy bracing members. The flange thickness tapers at an angle of 5° or 8°.

(d) Rolled-steel channels (RSCs or [). The flush back-face of the RSCs makes them very suitable where an I-section would be awkward. The flange thickness tapers at an angle of 5°.

22.4 Castellated beams

Castellated beams are UBs which have been cut and then welded back together to form a deeper section (Fig. 22.4). This greatly increases the moment of inertia of the beam and improves the resistance to deflection, but it also reduces the web strength.

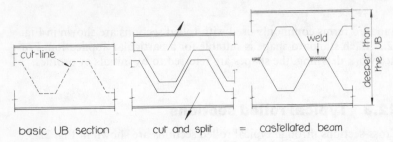

basic UB section   cut and split   = castellated beam

Fig. 22.4

Castellated beams are frequently used in long-span, lightly loaded structures such as warehouse roofs, where the critical design problem may be to restrict mid-span deflection and where the web loading is small.

Castellated universal columns and castellated joists are also available.

22.5 Angles (Angle or ∠) and Tees (Tee or T)

Angles are rolled to their finished profile (Fig. 22.5(a)). Some Tees are

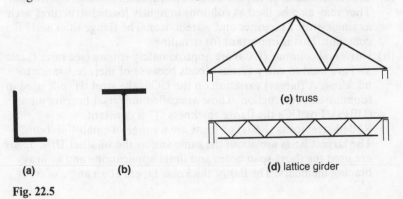

(a)   (b)   (c) truss   (d) lattice girder

Fig. 22.5

also rolled to the finished profile, but the majority are cut from UBs and UCs (b). Angles and Tees are used for general framing applications and in the fabrication of lightweight steel trusses and lattice girders (c, d).

22.6   Flats, bars and plates

Flats are flat rectangular sections in which the profile of the top, bottom and sides is controlled in the rolling mill. A very wide range of flats in different thicknesses and widths is available and they are used mainly for strapping, plating and stiffening in steelwork assemblies. A variation is the bulb flat which is J-shaped in cross-section. Bulb flats are commonly used as plate stiffeners in box-girder bridge construction.

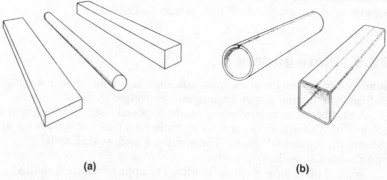

(a)   **(b)**

Fig. 22.6

Square and round bar is normally used in small assemblies and as diagonal members in lattice girders. A flat rectangular section and bars are shown in Fig. 22.6(a).

Plate is flat sheet formed by rollers top and bottom but with the edges left free as the rolling proceeds. Finished plate can be supplied in widths from 1 to 4 metres and thicknesses from 5 to 200 mm, although the maximum thickness of plate used in the construction industry is about 100 mm.

22.7   Structural hollow sections (SHS)

Structural hollow sections are made by drawing a flat steel strip through dies which form it into a tube. They may be either circular hollow section (CHS) or rectangular hollow sections (RHS). The tube is closed by a continuous exterior and interior weld (Fig. 22.6(b)). The exterior weld is skimmed flush, then the tube is heated and passed through a rolling mill where it is stretch-reduced to a circular or rectangular section. The rolling controls both the size of the tube and the wall thickness.

Tube is often used for exposed roofs in buildings because the assembled structure can be made to look very attractive. It is also used as bracing in frames made from UB and UC sections.

The maintenance of structures made from hollow sections is relatively easy because there are no internal angles.

22.8   Availability

The sections described are suitable for the majority of building structures. Rollings of all sections take place every few weeks and the steel-fabricator can place an order for the sections required directly with the manufacturer. The fabricator can also buy many of the more popular sections 'off the shelf' from a steel-stockholder.

22.9   Plate girders

Some heavy civil engineering structures such as bridges, power stations and heavy industrial works, require much stronger sections.

A number of steel-fabricators produce special beams and columns to order (Fig. 22.7(a)). They are built up from steel plates which are set up to form the required 'I' shape. The joints are then welded, usually by an automatic welding machine.

Welded plate girders can be made up to approximately 4 metres deep and 1.3 metres wide using plate up to 100 mm thick. Within those limits the designer can choose the exact sizes required. There is no standard range as with rolled sections. The flanges on the beam shown are of different widths. This is a common detail in welded plate work and presents no problems to the fabricator.

22.10   Steel plate flooring

Embossed steel sheet called Durbar Floor Plate is produced for decking on walkways, cat walks, etc. (Fig. 22.7(b)). It is made in plate thicknesses from 4.5 to 12.5 mm and the surface has a raised pattern which forms a free-draining anti-slip surface.

The plate is designed to span between supporting frames to which it may be welded or screwed. Tables are available which list safe loads and spans for different thicknesses of plate.

Fig. 22.7 (a)

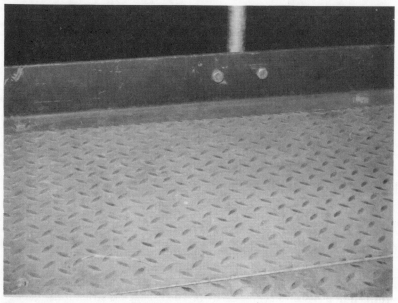

(b)

22.11 British Standards

Full dimensional details for rolled sections are set out in:

BS4: Part 1, which covers UBs, UCs, RSJs, RSCs and Tees cut from UBs and UCs.
BS4848: Part 2, which covers CHSs and RHSs.
BS4848: Part 4, which covers angles.
BS4848: Part 5, which covers bulb flats.

Structural sections have to be rolled or formed within tolerances, and these may affect the steelwork detailing, particularly at joints between members. Full details of permitted rolling tolerances are given in the British Standards. Figure 22.8 illustrates some of the tolerances applicable for the beams, columns and channels in BS4: Part 1. These are (a) dimensions, (b) web offset, and (c) out of square. There are also tolerances for straightness and weight.

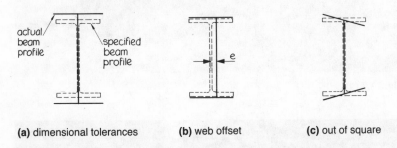

(a) dimensional tolerances   **(b)** web offset   **(c)** out of square

Fig. 22.8

Angles, hollow sections and bulb flats are also subject to dimensional and longitudinal distortion tolerances. Details are given in the appropriate British Standards.

Welded plate sections can be fabricated to any size within the limits described previously. Although they have to comply with the general requirements of BS5950/BS449, there is no British Standard exclusively for welded plate girders, beams and columns.

22.12 Cladding support

Lightweight steel purlins and sheeting rails are used to support the corrugated sheet claddings often used on steel-framed buildings.

The purlins and rails are formed from steel strip which is cold-formed to a Z-section (Fig. 22.1(b)). The Z-shape is structurally very efficient.

The sections are produced by a number of manufacturers as part of a proprietary cladding support system. The manufacturers' catalogues contain information about safe loads, purlin layouts and typical construction details. The steelwork detailer uses this information to produce scheme drawings.

22.13 Structural aluminium

Aluminium is also used for building and civil engineering structures.

Structural sections such as Is and tubes are made by the extrusion process to a range of profiles. They are used for portal-framed structures such as sports halls and warehouses, for lattice trusses and for tubular space framed roofs. Aluminium bridge parapets and vehicular guardrails (crash barriers) are also extensively used. Although aluminium does not corrode appreciably in normal atmospheric conditions, structures in which it is used with steel must be carefully detailed so that the two metals do not come into direct contact.

Contact between different metals in a normal atmosphere can result in sacrificial corrosion (section 26).

Section 23

The steel tables

23.1 Introduction

The rolled sections and tubes described in section 22 are available in a wide range of sizes. Detailed dimensions of the various sections are tabulated in the British Standards. However, steelwork designers and detailers do not normally use the British Standards directly. The steel industry publishes handbooks commonly known as 'steel tables' and 'safe load tables' which contain the section dimensions and structural properties listed in the Standards. They also include other information which is required for the design and detailing of structural steelwork such as: basic design formulae, strut and tie loads and the working strengths of bolts and welds.

Designers and detailers use the steel tables as a basic reference book.

23.2 Presentation

The steel tables include two sets of information for each group of sections (Fig. 23.1). The dimensional details are of particular interest to the steelwork detailer. Figure 23.2 shows an extract from part of the UB tables.

23.3 Universal beams

Universal beams are available in a wide range of sizes from 127×76 mm

full dimensional details, weights and cross-sectional areas ⟩ ⟨ section properties required for use in structural design – moments of inertia, radii of gyration, elastic and plastic moduli

Fig. 23.1

(depth × width) to 914 × 419 mm. In the steel tables they are grouped into a number of nominal sizes under the heading 'Serial Size'. The range of serial sizes is listed in the first column of the table. For each serial size, a range of section properties is produced by varying the overall dimensions and metal thickness. The steel tables list the exact depth D, width B, web thickness t and flange thickness T for each beam.

For example, the 457 × 152 mm serial size includes five UBs each with slightly different dimensions.

Figure 23.3 shows dimensioned profiles for the largest and smallest 457 × 152 mm UBs.

Varying the depth and flange thickness has the greatest effect on the section properties. The bending strength of beam (a) is more than one and a half times that of beam (b).

Varying the overall dimensions and metal thicknesses also produces beams of different weights. These are measured in kg per metre and are tabulated in the second column of the tables (Fig. 23.2). All beams of a particular serial size have effectively the same distance between the inside faces of the flanges; in the example shown 427.3 mm and 428 mm.

The tables also list the radius of the flange/web fillets, called the root radius r, the area of section which is measured in square centimetres, and data which are required for detailing connections between structural sections. This is explained in section 29.

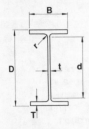

UNIVERSAL BEAMS

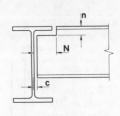

DIMENSIONS

Designation		Depth Of Section D	Width Of Section B	Thickness		Root Radius	Depth Between Fillets	Dimensions For Detailing			Area Of Section
Serial Size	Mass Per Metre			Web	Flange			End Clearance C	Notch		
				t	T	r	d		N	n	A
mm	kg	mm	mm	mm	mm	mm	mm	mm	mm	mm	cm²
457×152	82	465.1	153.5	10.7	18.9	10.2	406.9	7	82	30	104.5
	74	461.3	152.7	9.9	17.0	10.2	406.9	7	82	28	95.0
	67	457.2	151.9	9.1	15.0	10.2	406.9	7	82	26	85.4
	60	454.7	152.9	8.0	13.3	10.2	407.7	6	82	24	75.9
	52	449.8	152.4	7.6	10.9	10.2	407.7	6	82	22	66.5
406×178	74	412.8	179.7	9.7	16.0	10.2	360.5	7	96	28	95.0
	67	409.4	178.8	8.8	14.3	10.2	360.5	6	96	26	85.5
	60	406.4	177.8	7.8	12.8	10.2	360.5	6	96	24	76.0
	54	402.6	177.6	7.6	10.9	10.2	360.5	6	96	22	68.4
406×140	46	402.3	142.4	6.9	11.2	10.2	359.7	5	78	22	59.0
	39	397.3	141.8	6.3	8.6	10.2	359.7	5	78	20	49.4
356×171	67	364.0	173.2	9.1	15.7	10.2	312.3	7	94	26	85.4
	57	358.6	172.1	8.0	13.0	10.2	312.3	6	94	24	72.2
	51	355.6	171.5	7.3	11.5	10.2	312.3	6	94	22	64.6
	45	352.0	171.0	6.9	9.7	10.2	312.3	5	94	20	57.0
356×127	39	352.8	126.0	6.5	10.7	10.2	311.2	5	70	22	49.4
	33	348.5	125.4	5.9	8.5	10.2	311.2	5	70	20	41.8
305×165	54	310.9	166.8	7.7	13.7	8.9	265.7	6	90	24	68.4
	46	307.1	165.7	6.7	11.8	8.9	265.7	5	99	22	58.9
	40	303.8	165.1	6.1	10.2	8.9	265.7	5	90	20	51.5
305×127	48	310.4	125.2	8.9	14.0	8.9	264.6	6	70	24	60.8
	42	306.6	124.3	8.0	12.1	8.9	264.6	6	70	22	53.2
	37	303.8	123.5	7.2	10.7	8.9	264.6	6	70	20	47.5
305×102	33	312.7	102.4	6.6	10.8	7.6	275.9	5	58	20	41.8
	28	308.9	101.9	6.1	8.9	7.6	275.9	5	58	18	36.3
	25	304.8	101.6	5.8	6.8	7.6	275.9	5	58	16	31.4
254×146	43	259.6	147.3	7.3	12.7	7.6	218.9	6	82	22	55.1
	37	256.0	146.4	6.4	10.9	7.6	218.9	5	82	20	47.5
	31	251.5	146.1	6.1	8.6	7.6	218.9	5	82	18	40.0
254×102	28	260.4	102.1	6.4	10.0	7.6	225.1	5	58	18	36.2
	25	257.0	101.9	6.1	8.4	7.6	225.1	5	58	16	32.2
	22	254.0	101.6	5.8	6.8	7.6	225.1	5	58	16	28.4
203×133	30	206.8	133.8	6.3	9.6	7.6	172.3	5	74	18	38.0
	25	203.2	133.4	5.8	7.8	7.6	172.3	5	74	16	32.3
203×102	23	203.2	101.6	5.2	9.3	7.6	169.4	5	60	18	29.0
178×102	19	177.8	101.6	4.7	7.9	7.6	146.8	4	60	16	24.2
152×89	16	152.4	88.9	4.6	7.7	7.6	121.8	4	54	16	20.5
127×76	13	127.0	76.2	4.2	7.6	7.6	96.6	4	48	16	16.8

Fig. 23.2

23.4 Identification

All steel sections have to be identified on drawings. Universal beams are called up by their serial size and weight, so that the beams shown in Fig. 23.3 would be: (a) 457 × 152 × 82 kg UB, and (b) 457 × 152 × 52 kg UB.

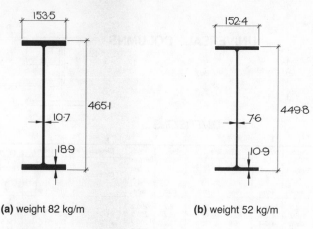

(a) weight 82 kg/m   **(b)** weight 52 kg/m

Fig. 23.3

23.5 Other sections

The steel tables covering UCs, RSJs, RSCs and castellated beams are similar to the UB tables. An extract from part of the UC tables is shown in Fig. 23.4.

23.6 Metric conversions

The serial sizes in the tables are direct metric conversions of the original imperial sizes. For example:

$$457 \times 191 \text{ mm} = 18'' \times 7\tfrac{1}{2}''$$
$$305 \times 127 \text{ mm} = 12'' \times 5''$$

These conversions apply to all rolled sections covered by BS4: Part 1: 1980.

There are internationally agreed standard metric dimensions for certain steel sections. Angles, hollow sections and bulb flats are now rolled to agreed metric sizes. The BS4: Part 1 sections will eventually be rolled to metric sizes.

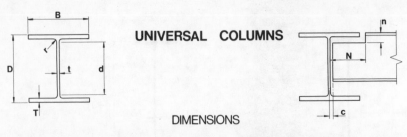

UNIVERSAL COLUMNS

DIMENSIONS

Designation		Depth Of Section D	Width Of Section B	Thickness		Root Radius	Depth Between Fillets	Dimensions For Detailing			Area of Section
Serial Size	Mass Per Metre			Web	Flange			End Clearance	Notch		
				t	T	r	d	C	N	n	A
mm	kg	mm	mm	mm	mm	mm	mm	mm	mm	mm	cm²
356x406	634	474.7	424.1	47.6	77.0	15.2	290.2	26	200	94	808
	551	455.7	418.5	42.0	67.5	15.2	290.2	23	200	84	702
	467	436.6	412.4	35.9	58.0	15.2	290.2	20	200	74	595
	393	419.1	407.0	30.6	49.2	15.2	290.2	17	200	66	501
	340	406.4	403.0	26.5	42.9	15.2	290.2	15	200	60	433
	287	393.7	399.0	22.6	36.5	15.2	290.2	13	200	52	366
	235	381.0	395.0	18.5	30.2	15.2	290.2	11	200	46	300
356x368	202	374.7	374.4	16.8	27.0	15.2	290.2	10	190	44	258
	177	368.3	372.1	14.5	23.8	15.2	290.2	9	190	40	226
	153	362.0	370.2	12.6	20.7	15.2	290.2	8	190	36	195
	129	355.6	368.3	10.7	17.5	15.2	290.2	7	190	34	165
305x305	283	365.3	321.8	26.9	44.1	15.2	246.6	15	158	60	360
	240	352.6	317.9	23.0	37.7	15.2	246.6	14	158	54	306
	198	339.9	314.1	19.2	31.4	15.2	246.6	12	158	48	252
	158	327.2	310.6	15.7	25.0	15.2	246.6	10	158	42	201
	137	320.5	308.7	13.8	21.7	15.2	246.6	9	158	38	175
	118	314.5	306.8	11.9	18.7	15.2	246.6	8	158	34	150
	97	307.8	304.8	9.9	15.4	15.2	246.6	7	158	32	123
254x254	167	289.1	264.5	19.2	31.7	12.7	200.3	12	134	46	212
	132	276.4	261.0	15.6	25.3	12.7	200.3	10	134	40	169
	107	266.7	258.3	13.0	20.5	12.7	200.3	9	134	34	137
	89	260.4	255.9	10.5	17.3	12.7	200.3	7	134	32	114
	73	254.0	254.0	8.6	14.2	12.7	200.3	6	134	26	92.9
203x203	86	222.3	208.8	13.0	20.5	10.2	160.9	9	108	32	110
	71	215.9	206.2	10.3	17.3	10.2	160.9	7	108	28	91.1
	60	209.6	205.2	9.3	14.2	10.2	160.9	7	108	26	75.8
	52	206.2	203.9	8.0	12.5	10.2	160.9	6	108	24	66.4
	46	203.2	203.2	7.3	11.0	10.2	160.9	6	108	22	58.8
152x152	37	161.8	154.4	8.1	11.5	7.6	123.5	6	84	20	47.4
	30	157.5	152.9	6.6	9.4	7.6	123.5	5	84	18	38.2
	23	152.4	152.4	6.1	6.8	7.6	123.5	5	84	16	29.8

Fig. 23.4

23.7 Angles

The dimensional details for rolled angle sections are also listed in the steel tables. 'Equal' and 'unequal' angles are rolled in sizes from

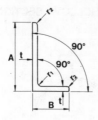

UNEQUAL ANGLES

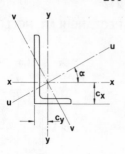

DIMENSIONS

Designation		Mass Per Metre	Radius		Area Of Section	Distance Centre Of Gravity	
Size	Thickness		Root	Toe			
A B	t		r1	r2		cx	cy
mm	mm	kg	mm	mm	cm²	cm	cm
200x150	18	47.1	15.0	4.8	60.0	6.33	3.85
	15	39.6	15.0	4.8	50.5	6.21	3.73
	12	32.0	15.0	4.8	40.8	6.08	3.61
200x100	15	33.7	15.0	4.8	43.0	7.16	2.22
	12	27.3	15.0	4.8	34.8	7.03	2.1
	10	23.0	15.0	4.8	29.2	6.93	2.01
150x90	15	26.6	12.0	4.8	33.9	5.21	2.23
	12	21.6	12.0	4.8	27.5	5.08	2.12
	10	18.2	12.0	4.8	23.2	5.0	2.04
150x75	15	24.8	11.0	4.8	31.6	5.53	1.81
	12	20.2	11.0	4.8	25.7	5.41	1.69
	10	17.0	11.0	4.8	21.6	5.32	1.61
125x75	12	17.8	11.0	4.8	22.7	4.31	1.84
	10	15.0	11.0	4.8	19.1	4.23	1.76
	8	12.2	11.0	4.8	15.5	4.14	1.68
100x75	12	15.4	10.0	4.8	19.7	3.27	2.03
	10	13.0	10.0	4.8	16.6	3.19	1.95
	8	10.6	10.0	4.8	13.5	3.1	1.87
100x65	10	12.3	10.0	4.8	15.6	3.36	1.63
	8	9.94	10.0	4.8	12.7	3.27	1.55
	7	8.77	10.0	4.8	11.2	3.23	1.51
80x60	8	8.34	8.0	4.8	10.6	2.55	1.56
	7	7.36	8.0	4.8	9.38	2.51	1.52
	6	6.37	8.0	4.8	8.11	2.47	1.48
75x50	8	7.39	7.0	2.4	9.41	2.52	1.29
	6	5.65	7.0	2.4	7.19	2.44	1.21
65x50	8	6.75	6.0	2.4	8.6	2.11	1.37
	6	5.16	6.0	2.4	6.58	2.04	1.29
	5	4.35	6.0	2.4	5.54	1.99	1.25
60x30	6	3.99	6.0	2.4	5.08	2.2	0.72
	5	3.37	6.0	2.4	4.29	2.15	0.68
40x25	4	1.93	4.0	2.4	2.46	1.36	0.62

Fig. 23.5

25×25 mm to 250×250 mm. An extract from the tables for unequal angles is shown in Fig. 23.5. The first two columns give the designation for the angle. This is similar to the serial size used for UBs but for angles the designated size is also the actual size. The designated size is expressed as the two leg lengths and the section thickness. For example:

$$150 \times 90 \times 10 \text{ mm} \angle \text{ (or angle)}$$
$$100 \times 75 \times 12 \text{ mm angle (or } \angle\text{)}$$

212

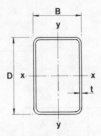

DIMENSIONS

DIMENSIONS

Designation		Mass Per Metre	Area Of Section
Size D B mm	Thickness t mm	kg	A cm²
120×60	3.6	9.72	12.4
	5.0	13.3	16.9
	6.3	16.4	20.9
	8.0+	20.4	25.9
120×80	5.0	14.8	18.9
	6.3	18.4	23.4
	8.0	22.9	29.1
	10.0	27.9	35.5
150×100	5.0	18.7	23.9
	6.3	23.3	29.7
	8.0	29.1	37.1
	10.0	35.7	45.5
	12.5+}	43.6	55.5
160×80	5.0	18.0	22.9
	6.3	22.3	28.5
	8.0	27.9	35.5
	10.0	34.2	43.5
	12.5+}	41.6	53.0
200×100	5.0	22.7	28.9
	6.3	28.3	36.0
	8.0	35.4	45.1
	10.0	43.6	55.5
	12.5	53.4	68:0
	16.0	66.4	84.5
250×150	6.3	38.2	48.6
	8.0	48.0	61.1
	10.0	59.3	75.5
	12.5	73.0	93.0
	16.0	91.5	117

+ Sections marked thus are not
included in BS4848: Part 2
} Sections marked thus are rolled in
grade 43C only

Designation		Mass Per Metre	Area Of Section
Outside Diameter D mm	Thickness t mm	kg	A cm²
21.3	3.2	1.43	1.82
26.9	3.2	1.87	2.38
33.7	2.6	1.99	2.54
	3.2	2.41	3.07
	4.0	2.93	3.73
42.4	2.6	2.55	3.25
	3.2	3.09	3.94
	4.0	3.79	4.83
48.3	3.2	3.56	4.53
	4.0	4.37	5.57
	5.0	5.34	6.80
60.3	3.2	4.51	5.74
	4.0	5.55	7.07
	5.0	6.82	8.69
76.1	3.2	5.75	7.33
	4.0	7.11	9.06
	5.0	8.77	11.2
88.9	3.2	6.76	8.62
	4.0	8.38	10.7
	5.0	10.3	13.2
114.3	3.6	9.83	12.5
	5.0	13.5	17.2
	6.3	16.8	21.4
139.7	5.0	16.6	21.2
	6.3	20.7	26.4
	8.0	26.0	33.1
	10.0	32.0	40.7

Fig. 23.6

23.8 Structural hollow sections

Extracts from the rectangular and circular hollow section tables are
shown in Fig. 23.6. Rectangular and square hollow sections are

identified by their side lengths and wall thickness. Referring to the extract:

$$\text{rectangular hollow sections} = D \times B \times t$$
$$\text{square hollow sections (not shown)} = D \times D \times t$$

Circular hollow sections are identified by their outside diameter and wall thickness. Referring to the extract:

$$\text{circular hollow sections} = \text{dia.} \times t$$

23.9   Surface areas

Surface areas of the different sections are also listed in the steel tables. These are used for estimating quantities for corrosion-protection systems such as galvanising and painting.

In working up a steelwork bill of quantities, the different structural sections may be itemised by lengths in metres or weights in tonnes. The table gives a direct conversion to surface area in square metres for either.

23.10   Plate girders

Welded plate girders, beams and columns are produced to the individual requirements of the designer and are not available ready-made from a steel supplier. Steel tables, as such, are not published for these sections although individual fabricators do produce information tables for use by designers and detailers.

Section 24

Bolted joints

24.1 Introduction

Structural steelwork is built from individual elements which are fabricated in a workshop, transported to site and assembled into complete frames.

In the workshop the elements are prepared for site assembly, holes are drilled as required, the drilled end-plates and brackets are welded on. On site, the elements are normally assembled into frames using bolted connections.

This section describes the different types of bolts used and explains how they should be detailed.

24.2 Bolted connections

A simple bolted connection normally comprises a bolt, a (flat) washer and a nut. The washer is put under the component which is to be turned (usually the nut). It acts as a slip surface during assembly and helps to maintain the tightness of the completed connection. When bolted connections are used with tapered sections such as RSJ flanges, a tapered washer is used against the angled surface to present a perpendicular seating for the nut or bolt. This may be used in addition to the flat washer. The podger spanner is the basic tool used for hand-tightening bolted connections. Torque spanners are also used with certain types of bolt.

The basic bolt, nut, washer fixing will be referred to as a 'fastener'.

24.3 Types of bolt

Several types of bolted connections are in common use. Their structural characteristics vary, and each type is appropriate to a particular application. Requirements for the components are specified in British Standards.

Mild steel, general-purpose fasteners are specified in:

BS4190: ISO Metric Black Hexagonal Bolts, Screws and Nuts;
BS4320: Metal Washers for General Engineering Purposes.

They are suitable for light and moderately loaded structures.

ISO metric precision hexagonal bolts are specified in:

BS3692: ISO Metric Precision Hexagonal Bolts, Screws and Nuts.

They are generally used as a high-strength version of the general-purpose (BS4190) bolted connectors.

High-strength friction grip (HSFG) connectors are a particular type of fixing which are tightened to a controlled torque. They are specified in:

BS4395 (3 parts): High strength friction grip bolts and associated nuts and washers for structural engineering.

24.4 Bolt strengths

The different bolt strengths are expressed in a decimal number form, for example grade 4.6. The stress/strain variation for the metals is similar to the graph shown in Fig. 21.3. In a bolt grading, the first number (4) is one-tenth of the ultimate tensile strength of the metal measured in kgf/mm^2. The second number (6) indicates that the yield stress is 60 per cent of the tensile strength. Bolts are matched with nuts of the same or a higher strength. Nuts are identified by the tensile strength, e.g. grade 8.

The shape of the stress/strain curve varies for different types of steel but the method of defining bolt strength is common to all. It is essential that each individual nut, bolt and washer can be identified by type and strength. This is achieved either by identifying markers embossed on each item or, in the case of washers, by shape.

24.5 Thread profile

The metric bolts used in the construction industry are made with coarse metric threads. The profile is based on BS3643: ISO Metric screw threads. This is in accordance with ISO recommendations and the same basic form is used for (imperial) unified threads.

When bolts are shown on drawings, it is not necessary to draw the threads to a saw-tooth profile. Instead, parallel lines are drawn representing the major and minor diameters of the thread.

24.6 General-purpose bolts

ISO metric nuts and bolts to BS4190 are the general-purpose fasteners used in structural steelwork. They have relatively low strength and are best suited to structures where the loads are light or moderate. The British Standard specifies three grades, 4.6, 4.8, and 6.9, but of these only grade 4.6 is normally used in structural steelwork.

Grade 4.6 bolts are hot-forged and they are commonly referred to as 'black-bolts'. The black coloration is not an applied finish but a by-product of the manufacturing process which is harmless and may be left on the metal. Similar but higher-strength bolts (grade 8.8) are produced with either a black or bright metal finish which means that the term 'black-bolt' is not precise enough to completely define grade 4.6 fasteners. They are best referred to by their grade strength.

The bolts are used with plain washers specified in BS4320. The bolts are fitted into clearance holes which are nominally 2 or 3 mm larger than the bolt (Fig. 24.1(a)) and the nuts are tightened by hand using an ordinary (podger) spanner. If the bolted joint carries shear forces, it is assumed that, under load, the connected elements slip relative to one another until the bolts bear on the sides of the holes (b). The load is transferred across the joint entirely by the bolts. They must be able to carry the load as a shear force across the shank, and a bearing force on the plates. For example, the shear capacity of a 20 mm diameter grade 4.6 bolt is 39.2 kN (3.92 tonnes). The bearing capacity of the connection is based on the surface area of the bolt which is assumed to be in contact with the hole, taken to be the plate thickness times the bolt diameter. The bearing strengths of grade 43 and 50 plate are 460 N/mm^2 and 550 N/mm^2 respectively.

Bolts are produced to a range of threaded lengths, and the threaded section is likely to project into the grip length. In this case, only the crest

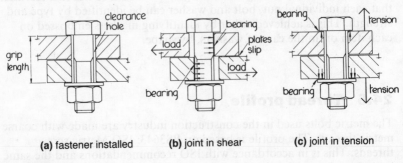

(a) fastener installed   (b) joint in shear   (c) joint in tension

Fig. 24.1

of the thread acts in bearing on the side of the hole, but the bearing capacity is not reduced; in fact it is slightly higher than for the plain, unthreaded shank. The thread may also cross the slip-plane between the connected plate, and so the design shear strengths for the bolts which are given in the steel tables and elsewhere are calculated on the minor diameter of the threaded section. If the joint carries tension forces (c), these are carried in bearing under the bolt head and nut/washer, and as tension in the bolt shank. If the joint carries both shear and tension, the separate reactions described will act in combination.

BS4190 details nuts and bolts from 5 to 68 mm in diameter and they increase in size mostly in 2 or 3 mm increments. However, the bolts are normally used for the site assembly of structural steelwork, often in difficult working conditions, where it is impractical to use very small fasteners even if they are theoretically adequate for the loads carried. Bolt diameters appropriate to construction are:

(M12)   M16   M20   (M22)   M24   (M27)   M30   (M33)   M36

Brackets indicate non-preferred sizes.

It is recommended that M16s should be the smallest bolts detailed for joints in structural steel frames and M12s the smallest for minor steel elements such as lintels in house conversions. Similarly, the largest bolts used should be M30s or M36s.

Detailed dimensions for grade 4.6 nuts, bolts and washers are given in the Standard which also includes a manufacturer's recommended range of lengths for bolt diameters from M12 to M24. Part of the range is shown in Fig. 24.2, which is taken from table 23 of BS4190: 1967. Certain popular sizes and lengths of bolt may be held by stockists; the others can normally be manufactured to order.

Manufacturer's recommended lengths for grade 4.6 hexagon head bolts																
	Length (mm)															
dia	35	40	45	50	55	60	65	70	75	80	90	100	110	120	130	140	150
M12	x	x	x	x	x	x	x	x	x	x	x	x	x	x		x	
M16	o	o	o	xo	xo	xo	xo	xo	xo	xo	xo	xo	x	x	x	x	
M20		o	o	o	o	xo	xo	xo	xo	xo	xo	xo	x	x	x	x	x
M24						o		xo		xo	xo	xo	x	x		x	

x=standard thread length, o=short thread length

Fig. 24.2

The standard thread length for the bolts is determined from Fig. 24.3, which is taken from table 4 of BS4190: 1967.

In an assembled connection, the standard thread lengths may project into the grip-length. This is not usually a problem, but if there is a specific reason why the thread projection should be reduced, an alternative short thread length of $1\frac{1}{2}d$ may be specified.

Nominal length of the bolt	Thread Length (bolt diameter d)
Upto and including 125mm	2d + 6mm
Over 125mm upto and including 200mm	2d + 12mm
Over 200mm	2d + 25mm

Fig. 24.3

24.7 High-tensile bolts

ISO metric high tensile bolts and nuts made to BS3692 may be cold-forged, hot-forged or turned from bar. The bolts and screws are made to strength grade 8.8 and the nuts to grade 8. The bolts are used in clearance holes and are designed to work in bearing in exactly the same way as the grade 4.6 bolts described previously. The dimensional tolerances applied to grade 8.8 connectors are finer than for grade 4.6, but for detailing, the principal dimensions (shank diameter, head sizes, etc.) may be assumed to be the same.

In designing and detailing fasteners, it should be possible to achieve the same strength with fewer 8.8 than 4.6 bolts. However, precision turning and other factors make grade 8.8 bolts more expensive, and they are not always the most suitable fastener. Both grades of fastener are commonly used. Manufacturers also produce a range of bolts in high-tensile grade 8.8 materials, made to the wider tolerance grade 4.6 profiles.

General-purpose washers to BS4320 should be used with grade 8.8 bolts and nuts.

24.8 Applications

Grade 4.6 and 8.8 bolts are designed to be tightened by hand, and when loaded, the connected plates slip relative to one another allowing the bolts to take the load in bearing. Load reversals and other dynamic effects would be carried directly by the bolts, and loosening or fatigue could occur. For this reason they should not be used in structures which are liable to dynamic loading.

24.9 Friction grip bolts

High-strength friction grip (HSFG) bolts are designed for use in connections where vibration or stress reversal may occur. They are used extensively in bridge construction. A joint made with HSFG bolts works differently from the bearing-bolt connections described previously. The bolts are fitted into clearance holes and are tightened to a predetermined value using a torque spanner or other load-indicating device. This controlled tightening causes compression forces between the connected plates (Fig. 24.4(a)). Where the joint is loaded in shear, the forces are carried by friction between the plates and the connection remains rigid.

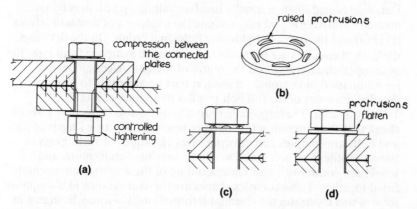

Fig. 24.4

The shear strength of the connection depends on the tension in the bolts (and therefore the compression between the plates) and the coefficient of friction (the *slip factor*) between the contact surfaces. This ranges from 0.5 for grit- or shot-blasted steel, to 0.05 for steel finished with red-lead paint. Galvanising, painting and surface grinding all reduce the slip factor and make a less efficient joint. The normal condition specified is a clean, lightly rusted surface, for which the slip factor is taken to be 0.45.

On the steelwork detail drawings, the joint-contact areas should be identified for protection from painting or galvanising so that connections are made between bare metal. The order in which bolts in a group are tightened is important, and is normally covered by notes on the drawings or as a separate schedule.

Friction grip bolts are produced to the requirements of BS4395, which is in three parts covering general and higher-grade bolts. Part 1 covers general-grade bolts which are the most commonly used. They are made from steels of similar quality to the grade 8.8 precision bolts described previously. They are referred to as 'general-grade HSFG bolts'.

The tension required in the bolts is achieved by controlled tightening. This may be done with a torque wrench, controlled hand-tightening, or using the 'CORONET' load-indicators shown in Fig. 24.4(b). These are flat, round washers with raised protrusions on one face. They are used in addition to a standard washer, and are normally fitted under the bolt head (c). As the nut is tightened and the tension in the bolt reaches the critical value, the segments are partially flattened. The remaining gap may be measured with feeler gauges (d) and checked against the design value.

24.10 Detailing bolts

Detailing bolted joints normally involves calling up fasteners by type, diameter and length, and may include the treatment of contact surfaces (HSFG) and the order of tightening bolts in a group. On the drawings, the bolts themselves are normally shown only in symbolic form typically as an open circle crossed on the centre of the hole. Occassionaly, for an unusual or complicated joint, it may be advisable to produce detailed drawings of the full bolt profiles to a large scale (1:5, 1:2, or full size). The large-scale illustration can also be used as a visual check that the connection fits together properly. The true shapes of nuts and bolts are complex and drawing exact elevations of such things as thread profiles and bolt heads would be very time-consuming, and would not contribute to an understanding of the way the bolt assembly fitted together. These complex shapes are therefore shown in a simplified form which illustrates the essential information. They may be drawn in accordance with the recommendations of PD7300, 'Nuts and bolts: recommended ratios for schools and colleges'. Figure 24.5 shows typical details. The simplified proportions used give values very close to the true dimensions, but in situations where the sizes are critical, the exact values given in the appropriate British Standards should be checked.

The bolt head and nut are usually projected across the corners to

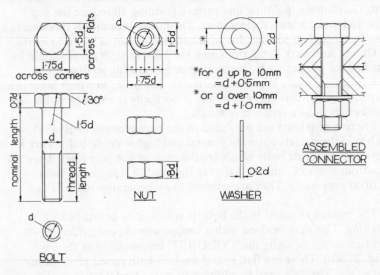

Fig. 24.5

show the wider elevation. If a bolt is shown from two directions at right angles, this wider elevation is often used for both views. When nuts and bolts are detailed in this way, it is important to produce neat, symmetric details. For small illustrations, the curved edges may best be drawn with a circle template or freehand, rather than attempting to use compasses. The broken line on the elevations of the bolt and nut indicate the start of the thread.

The assembled fastener (bolt, nut and washer) is shown as on a section taken at the centre-line of the hole. The plates are shown in section, but the fastener is drawn in elevation with the bolt passing through the nut and washer, which are both shown in full. Although a clearance hole is required, this need not be shown. When lock nuts are used they are placed between the standard nut and the washer and drawn 0.5d thick.

24.11   Bolt scheduling

On the steel fabrication drawings, the bolts are normally identified only by diameter and type. Full details of the complete fastener are scheduled as separate 'bolt lists' which accompany the drawings. Figure 24.6 shows the layout of a typical bolt list. The bolt list may be used to detail individual connections or as a summary sheet for a whole section of work.

BURNT MILL STEELWORK LTD

SITE BOLT LIST

Contract...

Quantity			Dia mm	Type	Grip	Length mm	Location	Washers	
Net	Spare	Gross						Flat	Taper

Fig. 24.6

The headings are:

Net—the number of bolts required in a connection or group of connections.
Spare and Gross—it is normal to allow for spare bolts against losses (about 5%).
Dia—diameter of the bolts (mm).
Type—strength grade and/or type.

222

Grip—lists of the thickness of the plates passed through.
Length—nominal length calculated from the thicknesses of the plates, washers and nuts, and the projection from the nut. The calculated length is rounded up to the nearest recommended length.
Location—brief description of the location in the structure including reference marks on the connected members.
Flat washer—the number of flat washers required.
Tapered washer—the number of tapered washers required.
The use of the schedule is illustrated by the example shown in Fig. 24.7.

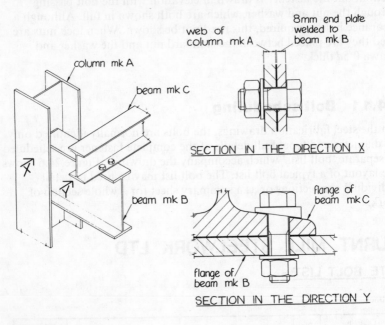

Fig. 24.7

Elements

The column mark A is a 305 × 305 × 97 kg UC.
The beam mark B is a 457 × 152 × 67 kg UB.
The beam mark C is a 203 × 152 RSJ.

Connections

Beam mark B to column mark A = 8 no. M16 grade 4.6 bolts.
Beam mark C to column mark B = 4 no. M16 grade 4.6 bolts.

It is assumed that the structure contains 10 no. beam-to-column and 6 no. beam-to-beam connections.

Bolt length calculation

Beam mark B to column mark A (mm)		Beam mark C to beam mark B (mm)	
Thro.		*Thro.*	
Welded end-plate	8.0	RSJ flange-mean thickness (from the steel tables)	16.5
Column web	9.9	UB flange (Fig. 23.2)	15.0
Length		*Length*	
Plates	17.9	Plates	31.5
Flat washer	3.0	Tapered washer	6.4
		Flat washer	3.0
Nut	13.0	Nut	13.0
Projection 2 pitches	4.0	Projection	4.0
	37.9		57.9
Round up to nearest 5 mm	40	Round up to nearest 5 mm	60

Use an M16 short thread:
thread length = $1\frac{1}{2}d = 24$ mm
(acceptable)

The projection is usually taken to be 2 (thread) pitches or 5 mm, whichever is the lesser.

The completed bolt list is shown in Fig. 24.8.

The use of the bolt list as a summary sheet (for ordering purposes) is shown in Fig. 24.9.

BURNT MILL STEELWORK LTD

SITE BOLT LIST

REDLAND DEVELOPMENTS   B6891

Contract NEW OFFICE BLOCK

Quantity			Dia mm	Type	Grip	Length mm	Location	Washers	
Net	Spare	Gross						Flat	Taper
					DWG B6891/15				
80	HEX	HD	M16	4·6	8,10	40	BEAM Mk B TO	1	
							COLUMN Mk C		
							(10 No THUS)		
36	HEX	HD	M16	4·6	16·5,15	60	BEAM Mk C TO	1	1
							BEAM Mk B		
							(6 No THUS)		

HEX HD = hexagon head

Fig. 24.8

Bolt lists for grade 8.8 and HSFG bolts are prepared in a similar way. To avoid confusion during the assembly of the structure, the number of different types, diameters and lengths of bolt should be kept to a minimum.

BURNT MILL STEELWORK LTD

SITE BOLT LIST

REDLAND DEVELOPMENTS

Contract...NEW OFFICE BLOCK........

No: B6891

Quantity			Dia mm	Type	Grip	Length mm	Location	Washers	
Net	Spare	Gross						Flat	Taper
			SUMMARY OF BOLTS						
			DWGS- B6891/12-16						
			ALL BOLTS AND NUTS-GRADE 4·6						
			FINISH: BOLTS AND NUTS TO BE HOT-DIPPED SPUN						
			GALVANISED TO BS729. ON COMPLETION, NUTS TO						
			BE TAPPED 0·4mm OVERSIZED AND THREADS						
			OILED PRIOR TO DISPATCH TO SITE.						
144	7	151	M16			40	HEX HD		
288	14	302	M16			50			
48	2	50	M16			60			
8	1	9	M20			50			
44	2	46	M20			70			
4	1	5	M20			100			
			FLAT AND TAPER WASHERS						
			FINISH:ALL WASHERS TO BE GALVANISED TO BS729						
		503	FLAT ROUND WASHERS FOR M16 BOLTS						
		60	"	"	"		" M20 "		
		104	8° TAPER		"		" M16 "		

Fig. 24.9

Section 25

Welding

25.1 Introduction

This data sheet describes common types of weld and how they should be shown on drawings.

25.2 Welding techniques

Structural steelwork is normally welded using an electric-arc process. The one most frequently used is called manual metal arc (MMA) welding. Figure 25.1(a) shows the arrangement of the equipment, which consists of a mains power source, leads, electrode and holder. The negative transformer lead is attached to the workpiece and when the electrode is touched onto the metal, a circuit is completed and an arc is struck, causing heat sufficient to melt the electrode and the workpiece in the area of the weld. The core of the electrode (b) is a wire which melts to form the weld.

The metal being welded is called the 'parent metal' and the welded surfaces are called 'fusion faces'. As the weld is laid, the parent metal melts and fuses with the electrode core (c). Welds must be formed to the correct profile and be protected from the atmosphere while they are being laid.

The electrode flux provides a gaseous shroud which protects the molten weld-metal from atmospheric contamination. It then forms a hard slag on the hot metal which gives further protection from

226

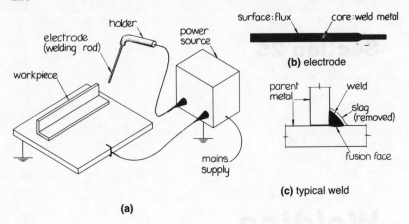

(b) electrode

(c) typical weld

(a)

Fig. 25.1

contamination and helps control the cooling rate. When the weld has cooled, the slag is removed.

The electrode is held in a spring-loaded grip and is consumed to form the weld. When it has been used up, the stub is discarded and another electrode fitted.

A range of electrodes is made for use with the different qualities of steel, and a properly made weld can be stronger than the parent metal.

There are also a number of automatic arc-welding techniques. Metal inert gas (MIG) welding uses a wire electrode which is fed continuously into the molten weld from a special head. The head also feeds an inert gas, either carbon dioxide or carbon dioxide and argon, which shrouds the weld and protects it from atmospheric contamination. Submerged arc (SA) welding also uses a continuous wire electrode but the flux is supplied as a powder which is automatically laid to completely enclose the arc. Automatic welding can also be carried out using continuous coated electrodes similar to those used in manual welding.

There are two basic types of welds: (a) fillet welds, (b) butt welds.

25.3 Fillet welds

Fillet welds (Fig. 25.2) are identified by the minimum leg length, for example: 8 mm, 12 mm; but the throat thickness is used for design purposes because that is the narrowest part. The weld face should be either flat or slightly convex (a). For a fillet weld made between two pieces of metal at right angles (b), it is normal to assume that:

$$\text{throat thickness} = 0.7 \times \text{leg length}$$

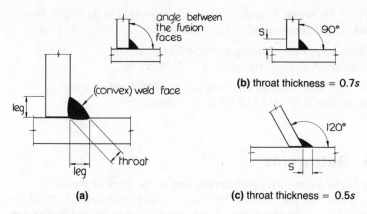

Fig. 25.2

As the angle between the fusion faces increases (c), the effective throat thickness reduces until at an angle of 120°:

$$\text{throat thickness} = 0.5 \times \text{leg length}$$

When the arc is struck at the start of a fillet weld, there is a short run-in before the full profile can be laid. Similarly, as the weld is finished, the profile tapers away over a short distance. These ends do not have the same strength as the full profile, and to allow for this, the effective length of an open-ended fillet weld is assumed to be $2s$ less than the overall length (Fig. 25.3(a)).

The minimum effective length of a fillet weld must be four times the leg length ($4s$) and when a weld is terminated at the end or side of a member, it should be returned around the corner a distance of not less than twice the size of the weld (b).

The design strength (p_w) of a fillet weld is dependent on the grade of steel, the electrode strength and the weld size. Typical values are 215 and

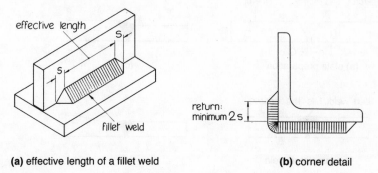

(a) effective length of a fillet weld

(b) corner detail

Fig. 25.3

255 N/mm² for grade 43 and 50 steels respectively. For example, for fillet welds made between plates at right angles:

6 mm fillet weld, design strength = 0.903 kN/mm run
8 mm fillet weld, design strength = 1.2 kN/mm run.

A 6 mm fillet weld with an effective length of 50 mm has a design strength of $50 \times 0.903 = 45.15$ kN (4.52 tonnes).

25.4 Butt welds

Butt welds lie substantially within the line of the faces of the parts joined. When two plates are to be butt welded, chamfers are formed on abutting edges (edge preparation) so that a groove is formed between them, and the weld fills the groove. Many different types of butt weld are used. The profiles are determined by the thickness and alignment of the metals being joined, and the strength required.

Figure 25.4 shows a single V butt weld. The parent metal is chamfered to form a 60° Vee (a). The root face R is 1 or 2 mm depending on the metal thickness, and the gap G is about 2 mm. When the weld is laid it fills the Vee and is finished to a convex surface (b).

When single V butt welds are used to connect plates of different thicknesses, if static loads only are applied to the weld, the detail (c) is used. If the weld carries dynamic loads which can cause fatigue, the thicker plate must be reduced (d), so that the weld is made between plates of the same thickness.

Figure 25.5 shows a single-bevel butt weld. The face of the intersecting plate is chamfered at 45°. The root face R is 1 or 2 mm, depending on the metal thickness, and the gap G is about 2 mm (a). When the weld is laid, it fills the Vee and is finished to a convex surface (b).

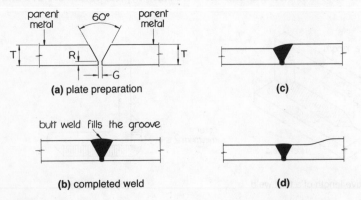

(a) plate preparation

(c)

butt weld fills the groove

(b) completed weld

(d)

Fig. 25.4

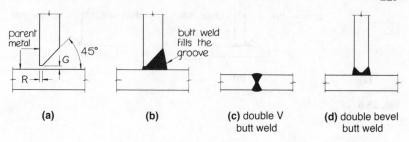

Fig. 25.5

Double V and double-bevel butt welds are also used (c, d).

The size of a butt weld is defined by the throat thickness, which is the thickness of the metals being joined. The strength is assumed to be the same as the parent metal, and when different thickness plates are joined, the strength is based on the thinner.

Edge preparations for butt welds are normally shown on the fabricator's drawings.

25.5 Notation

Welding details must be fully described on the drawings. Two methods of showing welds are used:

(a) British Standard BS499: Part 2 describes a system of symbols and arrows which can be used to define the type, size and location of welds.

(b) A popular system for detailing welds uses hatching and written descriptions. There is no standard set down for this technique, and the form of the descriptions may vary.

Although the BS system was introduced to encourage a standard method of detailing welds, the traditional method (b) is still used extensively, and so both methods are described.

25.6 Fillet weld notation

The BS method of describing fillet welds on drawings uses an indicator (Fig. 25.6(a)).

(a) The ▽ indicates a fillet weld.

(b) The 10 refers to the size of the weld in millimetres (the leg length). The figure is put on the left-hand side of the weld symbol.

(c) The arrowhead indicates the location of the weld. On a drawing it may not be convenient to point the arrowhead directly at the weld,

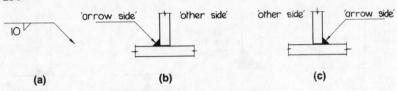

'arrow side'   'other side'   'other side'   'arrow side'

(a)   **(b)**   **(c)**

Fig. 25.6

which may be hidden from view. The notation system is extended to
allow for this. The definitions (b) and (c) apply and the location of
the weld in relation to the arrowhead is indicated by the position of
the symbol on the line (Fig. 25.7).

If the welds are of different sizes, the appropriate dimensions are
given beside each indicator.

The welds are assumed to be continuous. The line on which the
indicator triangle is located (called the reference line) is drawn
horizontal, and the triangle is drawn with the vertical face on the
left-hand side.

It is not necessary to show the sectioned weld; the arrow indicating
the location is sufficient.

(a) Symbol below the line:

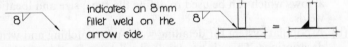

indicates an 8 mm
fillet weld on the
arrow side

(b) Symbol above the line:

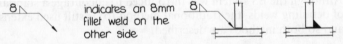

indicates an 8mm
fillet weld on the
other side

This would be used when it is not convenient to use the more direct method (a).

(c) A double symbol:

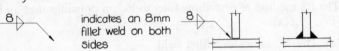

indicates an 8mm
fillet weld on both
sides

If the welds are of different sizes the appropriate dimensions are
given beside each indicator.

Fig. 25.7

25.7 Butt weld notation

Butt welds are specified on drawings using an indicator system similar to that used for fillet welds (Fig. 25.8).

For the 8 mm single Vee butt weld between in-line plates (Fig. 25.9) the weld symbol and size are shown below the reference line for a weld on the arrow side, and above the line for a weld on the other side. Double butt welds are shown by symbols above and below the line.

25.8 Intermittent welding

When intermittent lengths of weld are to be detailed, the lengths of weld and gap are indicated as shown in Fig. 25.10(a). The weld size in millimetres is given on the left-hand side of the symbol. The figures on the right-hand side detail the weld lengths (unbracketed) and gap lengths (bracketed). The gaps can become moisture and corrosion traps. Intermittent welding is not recommended in situations such as external and buried structures where corrosion could be a problem. Continuous welding is preferred.

A weld continuous around the periphery of a joint is specified by a circle on the indicator. For the plate/universal beam connection (Fig.

(a) single V butt weld   **(b)** single bevel butt weld

Fig. 25.8

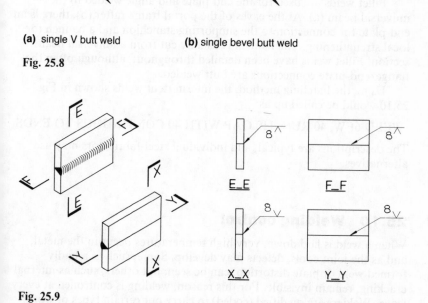

Fig. 25.9

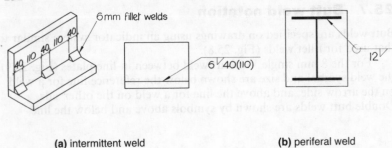

(a) intermittent weld   **(b)** periferal weld

Fig. 25.10

25.10(b)) the indicator specifies a 12 mm fillet weld around the perimeter of the beam.

The full range of weld indicators is given in BS499: Part 2.

25.9 Hatched notation

The hatching and description method for showing welds is illustrated in Fig. 25.11. The line and extent of the weld is hatched on the elevations and sections. Where a weld is seen in section this is shown. The welds are arrowed and a description is given.

Fillet welds are used for the end-plate and angle welded to the universal beam (a). At the eaves of the portal frame rafter (b), there is an end-plate for connection to the supporting stanchion and a haunch for local strengthening. The haunch is usually cut from a universal beam section. Fillet welds have been detailed throughout, although often the flange/end-plate connections are butt welded.

Using the hatching method, the intermittent welds shown in Fig. 25.10 would be called up as:

WELD 6FW, 40 RUN, 110 GAP WITH 40 CONTINUOUS TO ENDS

The descriptions are typical, and individual steel-fabricators may use alternatives.

25.10 Welding control

When a weld is laid down, very high temperatures occur in the metal, and as the joint cools, defects may develop. Some, such as a badly formed weld or plate distortion, can be seen; but others, such as internal cracking, remain invisible. For this reason, welding is controlled at every stage. Welders are qualified (coded) to carry out certain types of work.

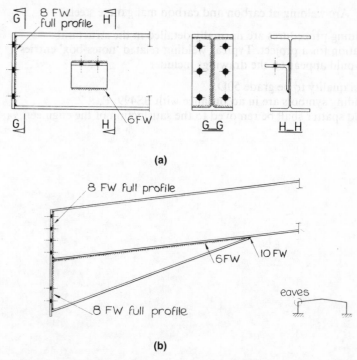

(a)

(b)

Fig. 25.11

On a contract, trials may be required to check the proposed welding procedures, and as the main works are carried out, test plates are made which can be tested for tensile strength, ductility and brittleness. Completed welds may be tested by non-destructive techniques including dyes which penetrate and reveal cracks, magnetic particle inspection, X- and gamma-rays and ultrasonics.

The requirements for production control, sampling and testing of welds are normally given in the steelwork specification and are not detailed on the drawings. The steel-fabricator prepares the welding procedures accordingly. These include weld design (a large weld may be laid down in ten or more passes, and the size and order must be decided), electrode type, current, voltage and position.

25.11 Drawing notes

British Standards which concern welding and weld detailing include:

BS499: Welding terms and symbols (parts 1 and 2).
BS639: Covered electrodes for the manual metal-arc welding of carbon and carbon manganese steels.

BS5135: Arc welding of carbon and carbon manganese steels.

Welding procedures are normally detailed in the structural specification for a project. Typical welding-related 'notes-box' entries which would appear on the drawings include:

1. Steel quality to be grade 50D.
2. Welding symbols are in accordance with BS499: Part 2.
3. Weld spatter shall be removed to the satisfaction of the engineer.

Section 26

Corrosion and fire protection

26.1 Introduction

This section describes methods of corrosion and fire protection for structural steelwork.

The carbon steels used for structural steelwork corrode in normal atmospheric conditions. When the sections are rolled, a chemical reaction takes place between the iron in the steel and oxygen in the atmosphere, forming an oxide skin called *millscale*. Millscale is very hard but also brittle and liable to crack. In normal moist atmospheric conditions a chemical reaction takes place in which the iron becomes oxidised to form hydrated ferric oxide (rust). The rusting takes place under the millscale, causing it to flake off. The process continues, causing pitting of the metal's surface and a consequent loss of cross-section and strength. It is not unusual in old unprotected steelwork for parts of the metal sections to be completely rusted away.

26.2 Protection systems

New steel should be protected from corrosion, and choosing the correct protection system is a very important part of steelwork detailing. Many factors must be considered. The environment and design life required from the system must be balanced against cost, including the eventual cost of maintenance. Painting the elements of a steel structure on a workshop floor is usually much cheaper than painting the assembled structure, when access towers and scaffolding become necessary, and the steelwork may eventually be hidden by finishes or become otherwise inaccessible for maintenance. Although many of the more durable protection systems are expensive, their extra cost may be insignificant compared with the eventual expense of the frequent maintenance necessary with an initially cheaper but less durable system.

Suitable corrosion protection systems may be selected from BS5493: Protective Coating of Iron and Steel Structures Against Corrosion, which describes a wide range of schemes, their use and specification. The Department of Transport 'Specification for Highway Works' describes protection systems and their application for highway structures.

A full description of the corrosion protection scheme should normally be given in the notes box on the drawings. If it is particularly complicated, or if the contractor is allowed to price several schemes, the drawing notes should list the alternatives.

Coating thicknesses are measured in microns (μm). A micron is one millionth of a metre (1000 μm = 1 mm).

26.3 Surface preparation

The first stage in all corrosion protection systems is thoroughly to clean the steel. Contaminants such as grease and oil must be removed using solvents and water rinsing. The metal may then be cleaned of millscale and rust using hand tools, acid pickling, flame or blast cleaning.

The finished quality is measured as the percentage of the surface which is completely cleaned, and is commonly identified as SA3 (100%), SA2½ (95%) and SA2 (80%).

In general, the more thoroughly the steel is cleaned, the longer the corrosion protection will last.

26.4 Galvanising

Galvanising is the coating of iron and steel with molten zinc. The zinc forms a metallurgical bond with the bare metal and the resulting protection system can give a long life before maintenance is required. The process is carried out by specialist sub-contractors working in factory conditions where quality control is good.

The standard surface preparation for galvanising is to pickle the steel in dilute hydrochloric or sulphuric acid and then apply a flux which absorbs any remaining impurities and keeps the metal clean until it is immersed into the bath of molten zinc. The zinc reacts with the steel to form a hard zinc/iron alloy layer, and as the steel is removed from the bath, a soft zinc skin forms (Fig. 26.1), which acts as a cushion against impact damage; the alloy layer gives abrasion resistance.

(soft) zinc

(hard) zinc-iron alloy

steel

Fig. 26.1

Galvanising is carried out to BS 729 Specification for hot-dip galvanising coatings on iron and steel articles.

The thickness of the coating is determined partly by the temperature of the molten zinc and the time of immersion, but mainly by the chemical composition of the steel and its surface roughness. When the steel is immersed in the zinc, the basic galvanised skin is established in about five minutes, and immersion for a much longer period would give only a small increase in thickness. The coating can be defined either by weight in g/m^2 or as a thickness in microns (μm).

BS729 specifies minimum coating weights:

For steel thicker than 5 mm—610 g/m^2 (= thickness 85 μm)
For steel from 2–5 mm—460 g/m^2 (= thickness 65 μm)
For steel less than 2 mm—335 g/m^2 (= thickness 47 μm)

26.5 Details for galvanising

Steelwork which is to be galvanised should be detailed to allow the molten zinc good access to all surfaces and it should also allow for free drainage as the steelwork is removed from the zinc bath. For example, in the joint shown in Fig. 26.2(a) the vertical angle is stopped short, leaving a gap for the free passage of zinc.

The web stiffeners (b) are chamfered, allowing a gap which prevents a build-up of zinc or zinc ash in an internal corner. The fabricated steelwork must have no enclosed compartments. In the detail (c), the circular hollow section (CHS) is capped at each end by channels. The tube is completely sealed and the high temperatures used for galvanising could cause a gaseous explosion in the compartment. The situation can be avoided by forming holes to allow the molten zinc in and the hot gases out, which normally means that the holes should be at the top and bottom of the tube.

Similarly, if two channels are placed back-to-back and the joint welded all round, a very thin compartment is formed (d). A hole drilled in the centre will allow the required access and ventilation. Holes should be a minimum of 8 mm diameter in small tubes and larger in other steelwork assemblies. They must be allowed for when the steel is detailed. After galvanising, the holes may be plugged for aesthetic reasons. Plugging is not essential since the metal is also coated on the inside.

238

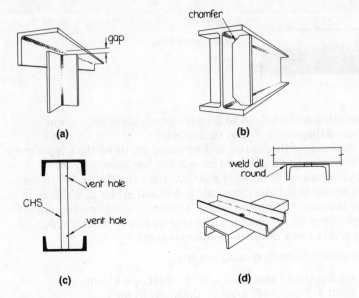

Fig. 26.2

When steelwork is fabricated, processes such as welding and
cold-bending can cause areas of local stress in the metal. The molten
zinc used in the galvanising process is at a temperature of about 450 °C
and while this is not high enough to alter the mechanical properties of
the steel, it can cause stress-relief at the welds and cold bends, resulting
in distortion. This can be a particular problem with lightweight
fabrications such as slender tubular structures. Distortion can also occur
due to lack of symmetry in the individual sections used and balance in
the fabricated steelwork. Symmetrical sections such as universal beams
and columns are less likely to distort than channels and angles. This
should be taken into consideration in designing and detailing the
steelwork.

26.6 Drawing notes

A typical set of notes-box entries are given for the galvanising of
steelwork to be erected in an exterior, exposed, polluted coastal
atmosphere (the classifications are taken from BS5493):

1. The corrosion protection scheme system is to be ref. SB1 or BS5493.
2. All steelwork shall be galvanised to BS729. The steel is to be:
   (a) cleaned of all grease, paint, welding slag and other
       contaminations;

(b) acid-pickled to remove all millscale and rust;

(c) hot-dip galvanised to a coating thickness of 85 μm.

3. Ventilation holes have been detailed for all sealed elements. The layout of the holes must be agreed by the galvaniser but if alternative ventilation is required, details of the revised positions and diameters must be submitted to the engineer for approval before work commences.

26.7 Advisory organisations

The Zinc Development Association and its affiliate Galvanisers Association offer an advisory service and publish such information on glavanising as the location and sizes of baths, budget quotations, design features, and the compatability of galvanised coatings with other materials.

26.8 Metal spray

Metal spraying uses aluminium or zinc to produce a protective coating. The metal, in the form of a powder or wire, is fed continuously through a special gun in which it is heated and converted into molten droplets which are then projected by compressed air to form an atomised stream. The droplets flatten on impact, forming a metallic coating of interconnecting plates. The coated metal is not heated appreciably by this process, and there is no risk of distortions occurring. The finished surface is usually protected with a sealer to give a smooth, attractive appearance.

Metal-sprayed coatings for corrosion protection should be specified to BS2569: Part 1: Protection of iron and steel by aluminium and zinc against atmospheric corrosion. Metal spray is also described in BS5493.

26.9 Painting

Paints are normally used in multiple-coat systems, starting with primers followed by under- and finishing coats. The types of paint used and the number of coats depend on the environment and the life to first maintenance required from the system. Encased steelwork in a heated building may require only one coat of a suitable primer, whereas steel exposed to a polluted atmosphere may need up to six coats of high-performance paint for full corrosion protection.

The thickness of the completed paint film influences the durability of the system and, in general, the thicker the film, the better the corrosion resistance. Within a system, each coat is applied to a controlled wet film thickness (wft) which becomes the dry film thickness (dft) when the paint dries. Dry film thicknesses of paint films are usually within the range 25 μm to 75 μm (high-build types) and it is possible to achieve thicknesses of 300–400 μm for specialist products.

Paint systems are detailed in BS5493.

Paint technology is very complicated and the British Steel Corporation advise on the suitability of different paint systems. The specifications described in the Standard are fulfilled by the product ranges of the paint manufacturers, who will also give technical advice.

26.10 Paint structure

The principal ingredients of a wet paint are a binder, solvent and pigment (Fig. 26.3(a)). The binder is the basic material forming the paint film and it determines many of the paint's characteristics and uses.

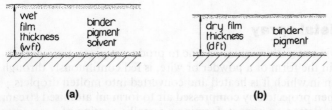

(a)   **(b)**

Fig. 26.3

Binders may be made from air-drying resins such as alkyds, or from naturally occurring elements such as linseed oil. Many are highly viscous (simply, viscosity is the reluctance of a liquid to flow) and the paint could not be applied without the addition of thinning agents (i.e. solvent).

Solvents may be added which dilute the paint and make it more workable. The mix is designed so that as the paint dries, the solvent evaporates freely, leaving a dry film of binder and pigment (b).

After application, the paint must dry in a reasonable time. Binders like linseed oil dry very slowly, and these can be chemically modified to speed up the drying process. Alternatively, the paint may dry by evaporation of the solvent (chlorinated rubbers) or by chemical reaction (epoxies). Two-pack epoxies are supplied in separate containers and mixed together immediately before use. This starts an irreversible chemical reaction and the paint must then be used within the 'pot life', after which it hardens.

The third basic ingredient of paint is pigment, which is particles of solid material added to the paint to make it opaque and add colour. Pigments may also improve water resistance, film strength and durability and improve 'chalking'. The materials used include metals (zinc, aluminium), metallic oxides (zinc and iron oxide) and other materials such as carbon black.

Many pigments are expensive, and to achieve the right balance in the ingredients, economic 'extenders' may be added which bulk up the paint and improve other characteristics.

26.11 Paint systems

A typical paint system suitable for an external location is shown in Fig. 26.4(a). The paints are chosen to prevent moisture from penetrating to the steel surface in the course of normal use, and, in the case of bare-metal damage, the system should be capable of resisting corrosion of the exposed metal.

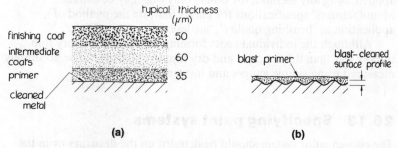

(a)   **(b)**

Fig. 26.4

Each paint performs a different function within the system. The finishing coat is the first line of defence against corrosion. It is fully exposed to the effects of ultraviolet radiation, rain, frosts and atmospheric pollution. The paint must have a tough, water-repellent surface. Gloss paints are normally designed with the minimum of pigment to achieve durability and opacity (typically 10–20% by volume of the dry film).

Intermediate coats are used to build the coating thickness and increase resistance to moisture penetration.

The primer is the first coat to be applied to the bare metal, and it is important that it should thoroughly cover the whole metal surface quickly and easily. Primers are designed as a direct protection for the metal, and they must also localise corrosion at any break in the paint film. There are several different types. Some contain inhibitive pigments such as red lead or zinc phosphate which chemically enshroud and stifle corrosion. Some two-pack epoxy primers are designed to adhere to the metal so strongly that local corrosion at a break in the paint film is unable to spread. Alternatively, the primer may contain pigments of a sacrificial metal. For example, zinc-rich primers usually contain a large quantity of zinc dust (typically 90% or more, by volume of the dry film).

Blast primers are applied to blast-cleaned steel in the workshops to prevent corrosion during the fabrication period. The depth and profile of the cleaned surface is determined by the type and size of the abrasive used (b).

When the corrosion protection system consists of galvanising followed by painting, an etch primer may be used on the galvanised metal. This etches the surface, giving a good key for subsequent coats.

26.12 Application

Paint can be applied by brush, roller or spray and the choice of system may be influenced by the method of application. Some paints can be applied using any method; for others only one may be suitable. Manufacturers' specifications for paints indicate the method of application as 'brushing quality', 'air-less spray quality', etc.

Although the individual coats forming the system are very thin, usually less than 0.1 mm, wet and dry film thicknesses can be accurately measured using simple gauges and instruments.

26.13 Specifying paint systems

The chosen paint system should be detailed on the drawings or in the specification. The paints are identified by type; proprietary names are not normally specified.

Within the paint system, successive coats of paint including stripe coats, should be different colours. This gives a simple means of checking that the full system is applied (stripe coats are additional coats applied along the edges where paint does not easily cover). It may be left to the contractor to choose colours except for the finishing coats which should be to the design team's requirements. They are normally specified to BS4800 which contains sample tiles of 100 different colours. Unless there are environmental or other restrictions, the design team does not normally specify the method of applying the paint.

For most steel structures, suitable paint systems may be chosen from BS5493. For highway structures, protection systems are also chosen for a particular environment, but details are given in part of the DTp Specification for Highway Works, which includes a standard paint system sheet that is used to record the design team's specification and the contractor's proposals for suitable paints.

An example of a paint system designed to BS5493 is shown in Fig. 26.5, which shows part of a steel frame supporting tanks in an industrial works. The frame is exposed to chemical pollution from the atmosphere, and the design life to first maintenance is 5–10 years. The steelwork is to be finished dark grey.

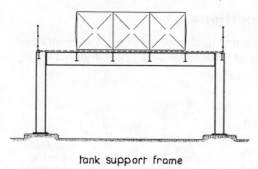

tank support frame

Fig. 26.5

A suitable paint scheme is chosen from BS5493 based on the environment, life to first maintenance and other factors. The Standard offers a range of primers, under- and finishing coats. The reference numbers are taken from the Standard.

Suitable notes-box entries are:

100. The corrosion protection scheme system for the structural steelwork shall be to BS5493, system reference SK2.
Surface preparation—blast clean to SA2½.
Primer (shop)—70 μm two-pack epoxy zinc phosphate chemical resistant (ref. KP1).
Undercoat (site)—100 μm two-pack epoxy micaceous iron oxide chemical resistant (ref. KUID).
Finish (site)—70 μm two-pack epoxy, rutile titanium dioxide (ref. KFIA), colour to BS4800, ref. 18 B 25.
Total dry film thickness 240 μm.
101. The paints, including those used for stripe coats, shall be of different colours.
102. Before steelwork fabrication commences, the contractor shall submit for approval full details of the proposed paint manufacturer and paint types including colours and method of application.
103. Primer shall be applied in the workshop. After site erection, damaged areas of paintwork shall be cleaned back to the engineer's approval, and repainted.

Paint systems for less aggressive environments may be more straightforward. Exposed steelwork in an office or shopping centre may be finished with a three-coat system of drying oil paints of total dry film thickness of less than 100 μm. Interior steelwork encased in architectural finishes may be given only one coat of high-build zinc phosphate primer 75 μm thick and steel in a totally dry interior environment may be left unprotected.

26.14 Special conditions

Bituminous paints are commonly used to protect buried steelwork and
can be applied satisfactorily as thick single coats, even in difficult site
conditions (Fig. 26.6(a)).

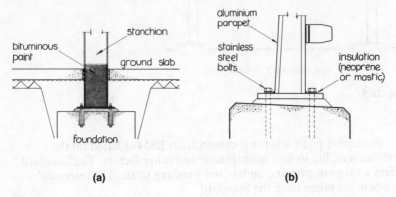

(a)   (b)

Fig. 26.6

Corrosion can also occur when dissimilar metals are placed in
contact in a moist atmosphere. An electro-potential difference
(measured in millivolts) occurs between them, and the moisture acts as
an electrolyte, causing sacrificial corrosion of one of the metals. The rate
of corrosion varies, depending on the particular metals in contact. In
some cases the problem is negligible, but for the combinations
commonly used in construction, the problem is often more serious. For
example, stainless steel holding-down bolts are used to secure the
aluminium parapet shown (b). If the metals touched, sacrificial
corrosion of the aluminium would occur. The problem is resolved by
insulating the metals using neoprene liners or mastic.

Full details of the protective measures must be shown on the
drawings.

26.15 Fire protection

It is a requirement of building regulations that, in the event of a fire, a
building will remain structurally stable for a sufficient length of time to
allow the occupants to escape safely (fire rating—measured in hours).
The regulations apply to multi-storey buildings and also to single-storey
buildings close to a site boundary.

In steel-framed buildings, the steel must be protected from the
intense heat which is generated in a fire. The critical temperature is

generally accepted as 550 °C, beyond which it rapidly loses strength. The steel may be protected by:

(a) Encasing the frame in cast-*in-situ* concrete reinforced with a light (wrapping) fabric reinforcement.
(b) Enclosure within brick or blockwork finishings.
(c) Enclosure within casings made from fire-resistant boarding such as plasterboard, mineral fibre and vermiculite/gypsum boards.
(d) The use of pre-formed claddings.
(e) Spray-on coatings based on mineral fibres or vermiculite.
(f) Intumescent coatings applied directly to the steel surface. When heated, the coating swells forming an insulating layer.

Normally, only the cast-*in-situ*-concrete system would be shown on the structural drawings. The others would be detailed by the architect.

Section 27

Hole spacing

27.1 Introduction

When structural steelwork connections are detailed it is necessary to follow certain guidelines regarding the spacing and positions of bolt holes.

This data sheet explains basic hole-spacing requirements.

27.2 Detailing holes

On structural steelwork drawings, open holes are normally shown shaded (Fig. 27.1). For bolted connections requiring clearance holes, these are made 2 or 3 mm larger than the nominal bolt diameter. The recommended sizes are:

bolt dia. (mm)   12 16 20 22 24 27 30 33 36
clearance hole (mm) 14 18 22 24 27 30 33 36 39

27.3 Edge distances

If bolt holes are drilled very close to the free edge of a steel plate or rolled section and the bolts are subjected to higher shear loads, they are liable to burst through the sides of the metal (Fig. 27.2(a)). To prevent this happening, holes should be located to give adequate edge (and end) cover. Minimum distances are given in Table 31 of BS5950: Part 1: 1985.

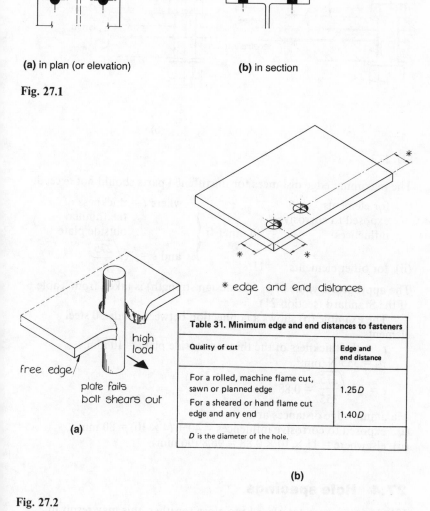

(a) in plan (or elevation) **(b)** in section

Fig. 27.1

* edge and end distances

Table 31. Minimum edge and end distances to fasteners	
Quality of cut	**Edge and end distance**
For a rolled, machine flame cut, sawn or planned edge	1.25D
For a sheared or hand flame cut edge and any end	1.40D
D is the diameter of the hole.	

free edge

high load

plate fails
bolt shears out

(a)

(b)

Fig. 27.2

These distances are measured to the centres of the holes and are dependent on the edge condition of the metal. Rolled, machine flame cut, sawn or planed edges are relatively uniform. Sheared or hand flame cut edges are slightly rough and an extra allowance is made to ensure that the basic minimum is achieved. The Standard also gives requirements for the maximum distance of a line of bolts from the nearest edge of a plate. They are intended to prevent excessive plate distortions when bolts are tightened (Fig. 27.3(a)).

248

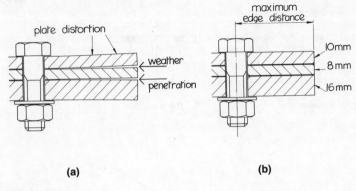

Fig. 27.3

The maximum edge distances for unstiffened parts should not exceed:

(i) for elements
     exposed to corrosive
     influences $= 40$ mm$+4t$

(ii) for other elements $=11t\epsilon$

where $t =$ thickness of
the thinner
outside plate

and $\epsilon = \sqrt{\dfrac{275}{p_y}}$

The appropriate value of p_y (the design strength) is taken from Table 6 of the Standard (section 21).

The example (b) shows a connection between grade 50 steel elements.

t = the thickness of the thinner outside plate = 10 mm.

$p_y = 355$ N/mm^2.

$\epsilon = \sqrt{\dfrac{275}{355}} = 0.88$

Maximum edge distances are:
(i)   exposed to corrosive influences $= 40 + (4 \times 10) = 80$ mm.
(ii) elsewhere $= 11 \times 10 \times 0.88 = 96.8$ (96 mm)

27.4 Hole spacings

If bolts in a group are placed too close together, this may result in congestion or overstress in the plates. The minimum allowable spacing (pitch) = $2\frac{1}{2} \times$ nominal bolt diameter. Also, the bolts must not be too far apart. For two adjacent bolts lying in the direction of stress (bearing bolts), the maximum pitch should not exceed:

(i) for elements
     exposed to corrosive
     influences $= 16t$ or 200 mm

(ii) for other elements $= 14t$

where $t =$ thickness of
the thinner
outside plate

27.5 Beams

Holes drilled in rolled sections should be positioned to maintain the minimum edge distances and hole spacings and prevent the bolted connection from fouling on the flange/web root radius (Fig. 27.4). Similar clearances should also be detailed near welds. The recommended spacings for holes in columns, beams and tees to BS4: Part 1, are given in the table in Fig. 27.5 based on the BCSA publication, *Metric Practice in Structural Steelwork*. Hole spacings shown in Figs. 27.7 and 27.8 are also taken from this publication.

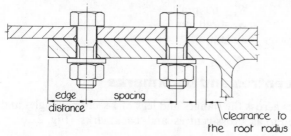

Fig. 27.4

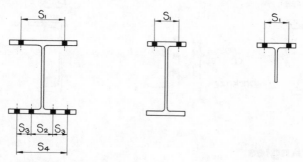

Nominal flange width	Spacing of holes				maximum dia. of fastener	'b' min
	S_1	S_2	S_3	S_4		
mm	mm	mm	mm	mm	mm	mm
419 to 368	140	140	75	290	24	358
330 and 305	140	120	60	240	24	308
330 and 305	140	120	60	240	20	295
292 to 203	140				24	208
191 to 165	90				24	158
152	90				20	145
146 to 114	70				20	125
102	54				12	89
89	50					
76	40					
64	34					
51	30					

Note that the actual flange width for a universal section may be less than the nominal size and that the difference may be significant in determining the maximum diameter.
The dimensions S_1 and S_2 have been selected for normal conditions but adjustments may be necessary for relatively large diameter fastenings or particularly heavy masses of serial size.
"b" min. This is the minimum width of flange to comply with Table 31 of BS 5950: Part 1.

Fig. 27.5

250

Examples

(a) 533 × 210 × 92 kg UB
Nominal flange width = 210 mm
From the table, S = 140 mm
Maximum bolt diameter = 24 mm

(b) 305 × 305 × 118 kg UC
Nominal flange width = 305 mm
From the table: for 2 holes S_1 = 140 mm
      for 4 holes S_2 = 120 mm
           S_3 = 60 mm
           S_4 = 240 mm
Maximum bolt diameter = 20 mm.

27.6   Cross-centres and backmarks

The spacing of holes across the flanges and legs of I-sections, angles and
channels are referred to as 'cross-centres' and 'backmarks' (Fig. 27.6).

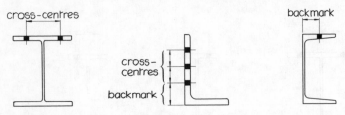

Fig. 27.6

27.7   Angles

The table shown in Fig. 27.7 gives recommended backmarks,
cross-centres and maximum bolt diameters for standard angles in
normal conditions. These spacings may not be appropriate for HSFG
fasteners which have larger nut and washer dimensions.

Examples

(a) 150 × 75 × 100 angle
From the table:
150 mm leg: S_2 = 55 mm, maximum bolt diameter = 20 mm
      S_3 = 55 mm, maximum bolt diameter = 20 mm
75 mm leg: S_1 = 45 mm, maximum bolt diameter = 20 mm

(b) 75 × 50 × 6 angle
From the table:
75 mm leg: S_1 = 45 mm, maximum bolt diameter = 20 mm
50 mm leg: S_1 = 28 mm, maximum bolt diameter = 12 mm

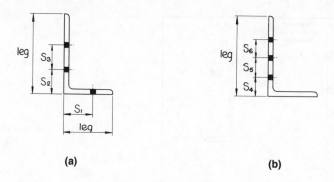

(a)   (b)

Nominal leg length	Spacing of holes						Maximum diameter of fastener		
	S_1	S_2	S_3	S_4	S_5	S_6	S_1	S_2 and S_3	S_4 S_5 and S_6
mm	mm	mm	mm	mm	mm	mm	mm	mm	mm
250				55	75	75			24
200		55	95	55	55	55		24	20
150		55	55					20	
125		45	50					20	
120		45	50					16	
100	55						24		
90	50						24		
80	45						20		
75	45						20		
70	40						20		
65	35						20		
60	35						16		
50	28						12		
45	25								
40	23								
30	20								
25	15								

Fig. 27.7

27.8 Channels

The table shown in Fig. 27.8 gives recommended backmarks and maximum bolt diameters for holes in the flanges of channel sections.

Example

229×89 RSC
From the table:
89 mm leg: $S_1 = 50$ mm, maximum bolt diameter = 20 mm

27.9 Slotted holes

It is occasionally necessary to detail slotted holes. For example, tolerances may be required in an assembled steelwork structure because

Nominal flange width	S_1	Maximum diameter of fastener	
mm	mm	mm	
102	55	20	
89	50	20	
76	45	20	
64	35	16	
51	30	10	
38	22		

Fig. 27.8

exact construction dimensions cannot be predetermined. This may occur with masonry support walls where construction tolerances are relatively coarse.

Slotted holes cannot be assumed to be fully effective in bearing. When a load is applied in the direction of the slot, the joint slips excessively until bearing occurs. Across the slot, bolts bear partly on a flat surface and do not have the necessary contact with the hole for full bearing to take place.

27.10 Example

The detailing of holes is illustrated by the beams and column example shown in Fig. 27.9; a similar example is used for bolt detailing in section

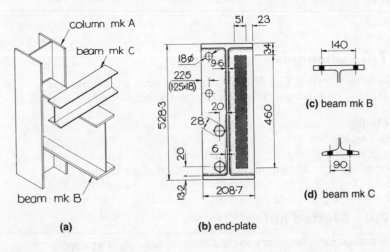

(a)

(b) end-plate

(c) beam mk B

(d) beam mk C

Fig. 27.9

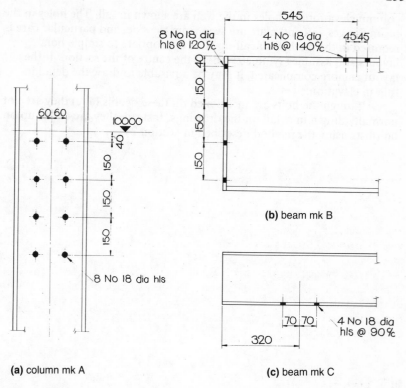

(a) column mk A

(b) beam mk B

(c) beam mk C

Fig. 27.10

24. The beam-to-column connection is made with 8 no. M16 bolts (grade 4.6). An 8 mm plate is welded to the end of the beam mark B (b) using a 6 mm fillet weld. The position of the 18 mm diameter clearance holes is determined from:

1. Minimum edge distance requirements.
2. The clearance of the bolt head (or washer) against the fillet weld.

The hatching indicates the area in which the holes can be drilled. The plate is conveniently made the same overall size as the beam, but would have to be larger if the holes cannot be positioned within the spacing requirements. If the plate is made wider, it must not foul on the internal radii of the column.

The beam-to-beam connection is made with 4 no. M16 bolts (grade 4.6) and suitable hole positions are chosen from the cross-centre tables. In this case S_1 for the RSJ is 90 mm and for the UB it is 140 mm. The 18 mm clearance holes are drilled on a 90×140 grid. Typical fabricator's details for these connections are shown in Fig. 27.10. The column and beams are drawn in elevation and the holes shown accordingly. On the

254

column elevation the holes in the web are shown in full. The holes in the beam flanges and end-plate are seen from the side, and particular care is required in the method of call-up. It is appropriate to simple bolt layouts and assumes symmetry about the centre of the section. If the layout is more complicated, it may be advisable to draw the detail in full, in elevation.

Although the bolts are mentioned on these details (a, c) they are not normally drawn in detail on the drawings. Instead, they are called up on bolt lists using the method described in section 24.

Section 28

Anchorage systems

28.1 Introduction

This section describes basic anchorage systems used for column bases and beam seatings.

The anchorage is the link between concrete foundations and structural steelwork. Foundations are cast *in situ*, and are subject to the normal tolerances of reinforced concrete construction. The steelwork is made in a workshop to close tolerances and must be erected on site with similar accuracy.

There are many types of anchorage systems for applications from lightweight building frames to heavy civil engineering structures such as power stations, and although the scale of construction may vary, the same problems must be considered and similar details used.

28.2 Column–base anchorage

A basic column–base anchorage system is shown in Fig. 28.1. The sequence of construction is:

(a) The holding-down (HD) bolts are cast into the concrete foundation. The top of the foundation is cast about 25–50 mm lower than the required column baseplate level.
(b) Steel shims are placed adjacent to the HD bolts and levelled so that the tops of the shims are at the column baseplate level.

256

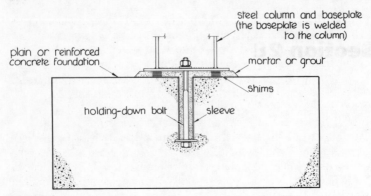

Fig. 28.1

(c) The column is placed on the shims and secured.
(d) The HD bolt sleeve and the gap between the foundation and the underside of the baseplate are grouted up.

28.3 Holding-down bolts

A typical holding-bolt detail is shown in Fig. 28.2(a). The bolts are cast upside-down in the concrete foundation. For construction they are normally bolted to a simple template which is fixed to the foundation shutter. The position tolerance on the bolts can only be as good as conventional shutter construction will allow. This could be up to ± 5 mm, which is more than the normal clearance on the column baseplate holes.

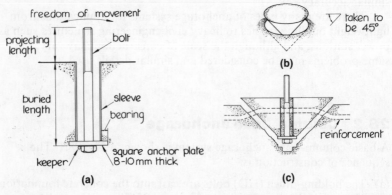

Fig. 28.2

The problem is overcome by forming a sleeve in the concrete which will allow some freedom of movement to the projecting end of the HD bolt. The sleeve can be formed using plastic or steel tubing which is left as a permanent liner. Alternatively, timber 'box-outs' or expanded polystyrene blocks can be used, but these must be removed before erecting the frame. The timber is pulled out, the expanded polystyrene is broken up and cleaned out with compressed air. The internal dimension of the sleeve is normally detailed to be about three times the bolt diameter.

Forces from the frame may cause shear and tension forces in the bolts. The diameter and buried length should be determined by calculation, but as a rough guide, 16, 20 and 24 mm diameter bolts are common and, in a small conventional building frame, the buried length will be about 250 mm. Tension in the bolts is transferred to the concrete through the anchor plates, which are normally made about 100 mm square, although calculations may be needed to determine the exact size. The force is resisted by the tensile strength of the concrete acting over the surface area of an inverted 'cone' around the bolt, and the buried length should be sufficient to create a cone large enough to prevent failure (b). The pull-out strength of the anchorage can be increased if steel reinforcement is fixed above the line of the anchor plates to pass through the potential cone of failure (c). The bolts must be prevented from turning when the nuts are fixed. This is done by welding a small steel 'keeper' to the anchor plate, or by tack-welding the bolt head directly to the plate.

The projecting length should be sufficient to allow for the column baseplate, the shims, the nuts and washers.

28.4 Stanchion bases

The stanchion base is formed by welding a steel plate to the end of the column (Fig. 28.3). The baseplate will be relatively thick: 12, 16 or 20 mm is quite normal for building frames and even thicker plates may be used in civil works.

Fig. 28.3 (a) (b)

The welds must be designed and then detailed on the drawings. A typical detail would be a continuous 8 mm fillet weld carried round the perimeter of the column. The continuous weld is not normally required for strength; short welds on the flanges should be sufficient to carry the loads at the column base but a continuous weld is a positive barrier against corrosion.

Four- and six-bolt anchorages (Fig. 28.3(a)) hold the column base rigid. This condition is described as 'fixed' and the bolts may have to carry large tension forces. The two-bolt anchorage (b) is described as 'pinned' and is more appropriate to lightly loaded structures.

28.5 Assembly

When the concrete foundation is cast, it is difficult to achieve a precise level for the top, which will have a tamped, slightly ribbed surface. Accurate levelling of the stanchion is achieved by casting the foundation low and then placing thin steel shims or folding wedges to the exact level of the baseplate. The stanchion is mounted on the shims and bolted down. Finally the anchorage is grouted up. Access to the bolt sleeves is difficult and they are normally filled with cement grout (cement:water 2:1 by weight) or by sanded grout (sand:cement:water 1:1:1 by weight). These are relatively fluid and can be poured into the hole. The gap under the baseplate is filled with sand/cement mortar (3:1:4 by weight) or fine concrete with a 10 mm maximum aggregate for large bases. The mortar is relatively dry and can be packed hard to completely fill the space. This is important because the vertical loads in the column are taken into the foundation through the mortar bed, not by the steel shims.

28.6 Drawing details

Figure 28.4 shows typical column anchorage details as they should appear on the structural drawings. The concrete foundation drawing (a) shows details of those elements of the anchorage which must be constructed with the foundation. The cast-in bolts and anchor plates are shown in the required positions. The steelwork assembly drawing includes details of the column base anchorage (b). This shows the vertical and horizontal location of the column and details the grouts, mortars, nuts and washers. The drawings do not include details of the levelling shims or of the method of forming the sleeves. These are both dealt with directly by the contractor.

The completed anchorage must be protected against corrosion. This may be done by painting with bituminous paint or by using corrosion-resistant fixings.

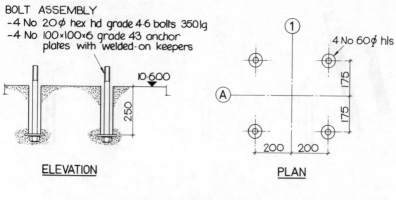

(a) anchorage detail

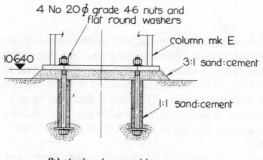

(b) steelwork assembly

Fig. 28.4

28.7 Drilled anchors

Although the cast-in holding-down bolt is in common use, there are alternatives. Drilled anchors, which are fixed into hardened concrete, can also be used and have certain advantages over the cast-in bolt.

A wide range of proprietary drilled anchors is available and they may be grouped into two categories; chemical anchors and expanding anchors. Figure 28.5(a) shows a typical chemical anchor. The sealed glass capsule contains resin and filler. It is inserted complete into a pre-drilled hole in the hardened concrete. The threaded bar is fixed into a power drill, placed in the hole behind the capsule and spun. This smashes the glass and the bar can be pushed to the bottom of the hole, the rotation ensuring that the resin and filler completely surround the bar. The resin hardens and the bar is locked in position with the end projecting to receive the column baseplate, nuts and washers. The hardening time for the resin can be controlled, although it can be as little as a few minutes.

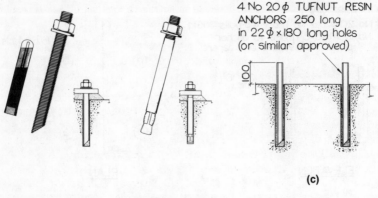

4 No 20 φ TUFNUT RESIN
ANCHORS 250 long
in 22 φ × 180 long holes
(or similar approved)

100

(c)

(a) a typical chemical
anchor

(b) a typical expanding
anchor

Fig. 28.5

In the typical expanding anchor (b), the steel stud is threaded at one end and has an inverted taper at the other. A loose expanding shell is fitted on the taper. The anchor is inserted, tapered end first, into a pre-drilled hole. It is locked into position when a nut is tightened on the projecting threaded section. The shell is designed to grip the sides of the hole. As the nut is tightened, the stud begins to withdraw and the taper expands the shell. The anchor locks, giving a fixing with a high pull-out resistance.

Accurate drilling of the holes is essential. Once fixed, these anchors cannot be moved to take up misalignments, unlike open-sleeved holding-down bolts.

Drilled anchors are proprietary products and the detailer must refer to manufacturers' catalogues to check that a proposed fixing is available. Figure 28.5(c) shows a typical anchor detail as it might appear on a drawing.

The product is specified, the hole is dimensioned and the projecting length of the threaded bar is given—this is critical. The phrase 'or similar approved' gives the contractor an opportunity to offer an alternative fixing which he may prefer for practical or economic reasons but it also gives the designer the right to reject its use.

28.8 Cast-in sockets

Cast-in sockets are threaded metal sleeves cast into the foundation concrete. Bolts can then be inserted and tightened down. Sockets are fixed—there is no freedom of movement to take up construction tolerances, and this can cause problems with multiple bolt fixings.

For the installation of a typical four-bolt detail, the group of sockets may be tack-welded to a simple steel harness. This prevents them from moving relative to one another during the casting operation.

28.9 Padstones

It is common in building structures to support steel beams on brick or block walls. Often these beams do not form part of a steel frame but are used in isolation; or they may be linked to a steel frame at one end and supported on a wall at the other. In either case there is no need for the sophisticated anchorage details already described for column bases.

For light loads and short spans a beam may bear directly on the masonry. As spans and loads increase, the beam could cause local crushing of the masonry and so a concrete padstone is cast onto the wall. The strong padstone takes the concentrated load from the beam and spreads it over a large area of the weaker masonry. The required size of the padstone will depend on the beam loads and the strength of the masonry wall. When blockwork is used it is convenient to detail the padstone to be the size of one block. A typical detail is shown in Fig. 28.6.

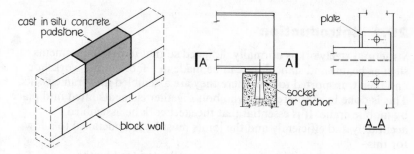

Fig. 28.6

If the beam is wide enough, the holes can be drilled directly through the bottom flange using standard cross-centres, but if the beam is narrow it is normal to weld on a rectangular plate with bolt holes drilled clear of the beam flanges. The plate does not have to be the same width as the padstone and if it is detailed 10–15 mm narrower on each side, it should not project beyond the face, even if the wall is slightly out of line. This may be important if the wall is to be plastered.

For this detail, proprietary threaded sockets may be cast into the padstone. Alternatively, drilled anchors may be used, but they should not be of the expanding type because the concrete padstone will not be very big, and the bursting forces which occur when the bolt is tightened would probably crack the concrete. Resin anchors do not have this problem.

Section 29

Steelwork connections

29.1 Introduction

Structural steelwork is normally designed so that individual elements such as beams and stanchions can be made in a fabricator's workshop and then transported to site where they are assembled into frameworks. This is done by steel-erectors who bolt together elements lifted into place by mobile crane. It is essential that the steelwork be assembled accurately and efficiently and the joints must be fully detailed to allow for this.

The joints in building frames are usually designed by the steel-fabricator who prepares separate elements and assembly drawings. To prepare the fabrication drawings for each element, the detailer must know what the assembled connections will look like. Details of the complete joints can usually be worked up as quick freehand sketches. The information is then used to prepare working drawings of the separate elements.

This section shows a range of basic joints used in steel frames.

29.2 Presentation

The illustrations show only the form of each joint. For a particular connection, the sizes of cleats, plates, welds and bolts must be calculated. Although indications have been given for the strength of welds (Section 25) and bolts (section 24), the full structural design of these connections is beyond the scope of this book.

The hatched method of showing welds is used to illustrate the typical joints but on fabrication drawings, either this or the arrow-and-triangle method may be used.

29.3 Beam/beam connections

Connections are described for beams which may be UBs, UCs, RSJs or channels.

End-plate (Fig. 29.1)

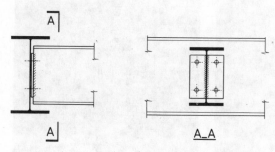

Fig. 29.1

The end-plate is welded to the secondary beam and the size is determined from the number, spacing and diameter of the bolts required, minimum edge distances, and the need for clearances against obstructions such as welds. This is a commonly used detail.

The plated secondary beam is a fixed length and for the assembly, shims may be required to take up fixing tolerances in the framework.

Cleats (Fig. 29.2)

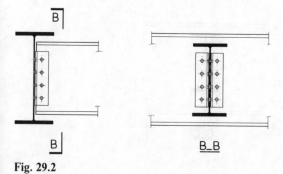

Fig. 29.2

264

The cleats are two angles bolted back-to-back through the web of the secondary beam. They project slightly, giving an end clearance. The connection is completed with bolts through the web of the main beam.

Occasionally, alignment problems occur in the assembly of steel frames, due to construction and fabrication tolerances. The 2 and 3 mm clearance holes used in the connection work like slotted holes, allowing small alignment adjustments to be made during the assembly. Alternatively, the cleats may be shop-welded to the secondary beam, the connection then works like the welded plate described previously.

Examples of end-plate and cleat fixings are shown in Fig. 29.3.

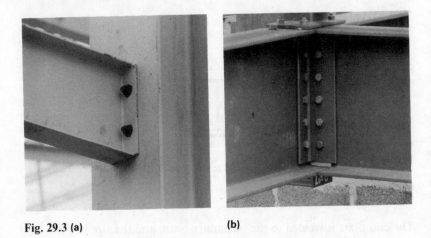

Fig. 29.3 (a)   **(b)**

Seating and side cleats (Fig. 29.4)

The angle seating cleat is welded to the main beam, and the side cleat to

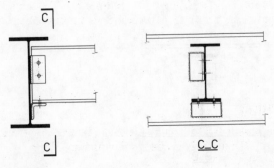

C   C_C

Fig. 29.4

the secondary beam. When the frame is assembled, the secondary beam is placed on the seating cleat which takes the weight, and the connecting bolts are fitted. The side cleat stabilises the connection.

A similar detail uses bolted connections throughout.

29.4 Pinned joints

The structural behaviour of a joint is controlled by the way in which the connection is detailed. If the joint is designed as 'pinned', it must be detailed to allow small rotations to occur. The end-plates and cleats should be thin (typically 8–10 mm). If they are rigidly attached (welded) they should be connected only to the web of the secondary beam. End clearances should be allowed, and the connecting bolts fitted into clearance holes (Fig. 29.5).

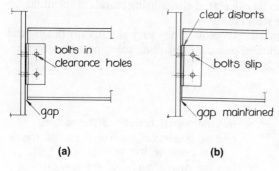

(a)   (b)

Fig. 29.5

Under load, free rotations take place at the joint because the bolts slip in the clearance holes, the thin plate distorts slightly and the beam flanges remain free and do not bear against the support.

29.5 Notch details

Beam-to-beam connections when the top flanges of the beams are at the same level (Fig. 29.6(a)). This detail is useful for beams supporting pre-cast concrete floors (b).

Profiled metal roof decking and steel floor plates can be fitted in the

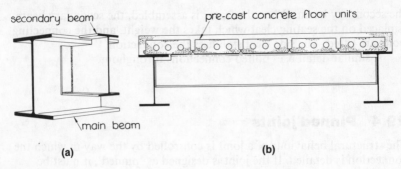

(a)   **(b)**

Fig. 29.6

same way. The secondary beam must be recessed to allow it to fit against the web of the main beam. This recess is referred to as a *notch* and the size is determined by the flange size of the main beam.

The dimensions can be calculated either using standard formulae or taken directly from the steel tables.

The formulae are tabulated below. They vary slightly for beams and columns, joists and channels, because the rolling tolerances vary for the different sections (Fig. 29.7).

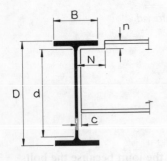

B is the overall breadth of the section.

c is the end clearance measured to the centre-line of the web.

D is the overall depth of the section.

d is the depth of the section between the corner radii.

N is the width of the notch measured to the face of the web.

n is the depth of the notch.

Fig. 29.7

For UBs and UCs

$$N = \frac{B - t}{2} + 10 \qquad \text{quoted to the nearest 2 mm above}$$

$$n = \frac{D - d}{2} \qquad \text{quoted to the nearest 2 mm above}$$

$$C = \frac{t}{2} + 2 \qquad \text{quoted to the nearest 1 mm}$$

For RSJs

$$N = \frac{D-t}{2} + 6 \qquad \text{quoted to the nearest 2 mm above}$$

$$n = \frac{D-d}{2} \qquad \text{quoted to the nearest 2 mm above}$$

$$C = \frac{t}{2} + 2 \qquad \text{quoted to the nearest 1 mm}$$

For RSCs

$$N = B - t + 6 \qquad \text{quoted to the nearest 2 mm above}$$

$$n = \frac{D-d}{2} \qquad \text{quoted to the nearest 2 mm above}$$

$$C = t + 2 \qquad \text{quoted to the nearest 1 mm}$$

Notch sizes are also tabulated in the steel tables (Figs 23.2, 23.4, etc.).

In both the formulae and tables, the notch width is measured to the face of the web, which includes an allowance for the end clearance. For detailing, the cut width of the notch must be adjusted accordingly.

Notched connections between beams can be made using the plate or cleat details described previously.

Beam-to-beam stepped connections (Fig. 29.8)

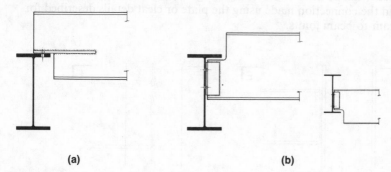

(a)   **(b)**

Fig. 29.8

(a) The bearing plate is slotted to fit into the web of the secondary beam.
(b) For the stepped connection, the notch width is determined from the standard formulae. The connection may also be detailed with the lower flange notched, making the secondary beam lower than the main beam (inset sketch).

Beam-to-beam butt joint (Fig. 29.9)

The joint shown is commonly used to form the ridge (apex) of a

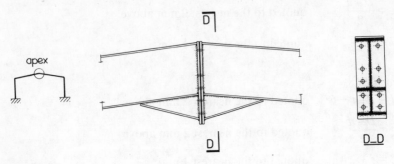

Fig. 29.9

pitched-roof frame. The haunches welded to the lower flanges increase the effectiveness of the connection. They may be cut from a UB section.

29.6 Beam/column connections (Fig. 29.10)

Beams can be fixed to either the flange (a) or the web (b) of a column, and the connection made using the plate or cleat details described for beam-to-beam joints.

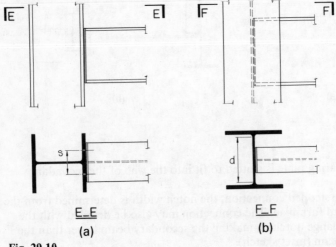

Fig. 29.10

For beam-to-column flange connections, the width between bolt holes (S) may be determined from the cross-centres (section 27). The width of plates and cleats used for beam-to-column web connections is restricted by the depth between fillets (d). An example of beam-to-column detailing is shown in Fig. 29.11.

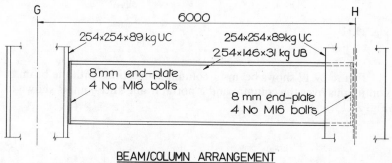

BEAM/COLUMN ARRANGEMENT

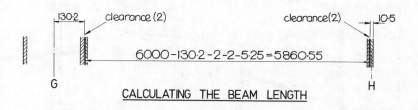

CALCULATING THE BEAM LENGTH

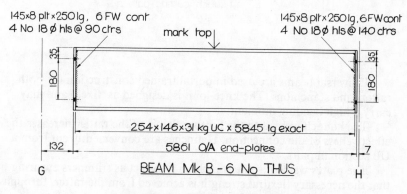

FABRICATION DETAILS FOR THE BEAM

Fig. 29.11

270

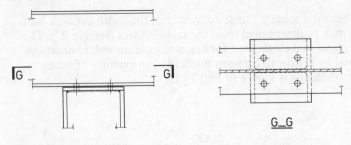

Fig. 29.12

Figure 29.12 shows beam-to-column connections with the beam continuous over the column, and a portal frame knee-joint is shown in Fig. 24.13.

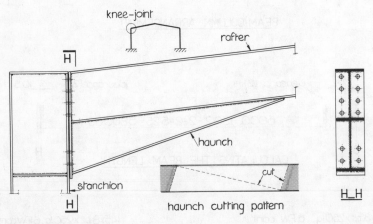

Fig. 29.13

Universal beams are used in portal framed construction for both rafters and stanchions. The knee-joint is designed as 'fixed' and may carry a high bending moment.

The haunch welded to the bottom flange of the rafter increases the effectiveness of the connection. Haunches are conveniently cut from a UB section in pairs.

The plates welded to the stanchion webs act as stiffeners ensuring that the necessary flexural strength is achieved from the rafter, through the joint and into the stanchion.

Figure 29.14(a) shows beams connected to the web and flange of a column. A portal frame knee-joint is also shown (b).

Fig. 29.14 (a)

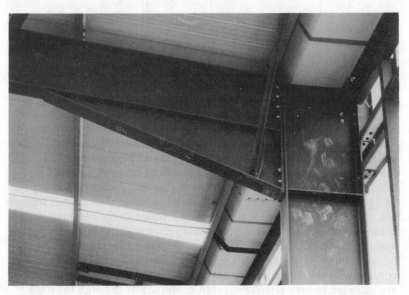

(b)

29.7 Column/column connections (Fig. 29.15)

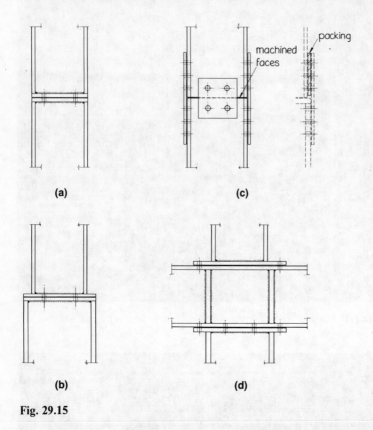

Fig. 29.15

Typical details are shown for splices between columns. In each case the load must be transmitted directly from column to column. For details (a) and (b), this is done through the welded end-plates. In detail (c) the load is transmitted through the butted column faces which must be machined to ensure continuous uniform contact. Vertical load may not be carried through the bolted web and flange cover plates. Steel sections are rolled within dimensional tolerances. Although the columns in the detail are shown the same size, they may have slightly different depths, and packing pieces are required to complete the joint. In the beam/column joint shown (d), the beam is continuous across the lower column with the upper column bolted to the top flange. Web stiffeners are included to give vertical strength continuously through the joint. Alternatively the column may be continuous through the joint with the beams bolted to the sides of the lower column.

29.8 Pin-jointed trusses

Figure 29.16 shows joint details for pin-jointed trusses. When designing pin-jointed frames it is assumed that the members carry axial loads only, and there are no joint bending moments. Members coming together at a joint should ideally be detailed so that their centroidal axes are coincident. This prevents eccentric loading of the joint which would cause bending moments.

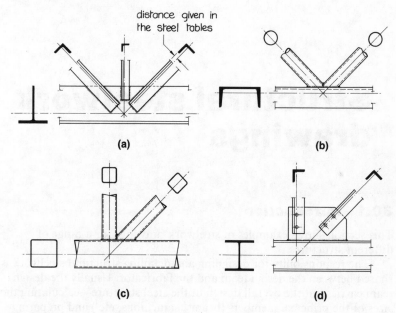

(a)

(b)

(c)

(d)

Fig. 29.16

The details show:

(a) A (shop) welded connection in which the angle section diagonal bracings are welded to the stem of the 'T'.
(b) A welded detail using a rolled-steel channel for the chord and circular hollow section bracings.
(c) A similar detail using a rectangular hollow section chord and bracings.
(d) A connection where the chord is a UC section and the angle bracings are bolted to a plate welded to the top flange. The line of the bolts passes through the member centroids. Alternatively, the angles could be welded to the plate.

Section 30

Structural steelwork drawings

30.1 Introduction

This section shows examples of steelwork drawings for a range of different structures.

The responsibility for designing and detailing structural steelwork is shared between the design team and the fabricator. Usually the design team carries out the overall design of the steel structure—calculating the sizes of the principal members (beams, stanchions, etc.) and preparing general arrangement drawings of the completed structure. These drawings form part of the tender documents.

The structural steelwork is normally made by a specialist steelwork fabricator, working as a sub-contractor to the main contractor for the project. Using the design team's drawings the fabricator prepares separate general arrangement and setting-out drawings which show the assembled steelwork. The fabricator also prepares drawings of the individual beams, stanchions, etc., which are used by the craftsmen and craftswomen making these elements in the workshop. For certain types of contract, the design team may prepare an outline scheme for the structural steelwork, but leave the detailed design to the fabricator. This arrangement is common in portal frame construction where fabricators who regularly carry out this type of work develop very economical designs.

The full-size drawing sheets are normally A1 (594 × 841 mm) to B1 (707 × 1000 mm) and the drawings include extensive detail which it is not possible to illustrate in a book of this size. However, the illustrations do show the typical content and style of the full-sized drawings.

It is not possible to predict exactly how many drawings will be required for a particular project, but as a rough indication, a typical steel frame for a moderate-sized building, for example Fig. 21.1(b), may require up to five design-team drawings and between ten and twenty fabricator's drawings.

30.2 Portal frame

Drawings are shown for a typical portal framed building of the type commonly used for warehouses and light industrial premises. Details are shown for:

(i)   The fabricator's general arrangement drawing—Fig. 30.1.
(ii) The fabricator's element drawing for a rafter—Fig. 30.2.

A typical design team general arrangement drawing for a structure of this type is shown in Fig. 21.5.

The fabricator's general arrangement drawing

This drawing shows a plan, side and end elevations of the complete structure.

These drawings are often set out on the grid system used for the design team's details. Primary dimensions are shown on the plan and elevations. The elements, rafters, stanchions and so on are identified by groups and are given mark numbers which are carried over to the individual element drawings.

The fabricator's element drawing for a rafter

This drawing shows:

(a) An identifying mark number which can be cross-referred to the general arrangement drawing.
(b) An elevation of the completed rafter which shows the horizontal, vertical and slope dimensions of the rafter and identifies the components—the UB section rafter, eaves and apex haunches, end-plates, gussets and the cleats required for the cladding purlins.

Separate elevations are not normally necessary for simple rectangular end-plates which have a symmetrical arrangement of holes. Dimensions given on the side elevation plus appropriate notes detailing plate sizes and cross-centres for the holes are sufficient. However, elevations are required to show hole layouts in more complicated shaped plates.

Fabricator's workshop drawings are often not drawn to scale. Instead they are drawn to emphasise areas of particular interest. On the rafter drawing, the ends and the splice are shown roughly to scale and the long lengths of uniform section between are drawn to a reduced

276

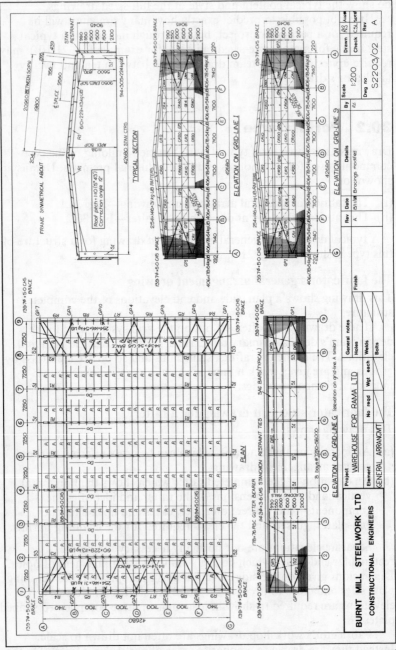

Fig. 30.1

277

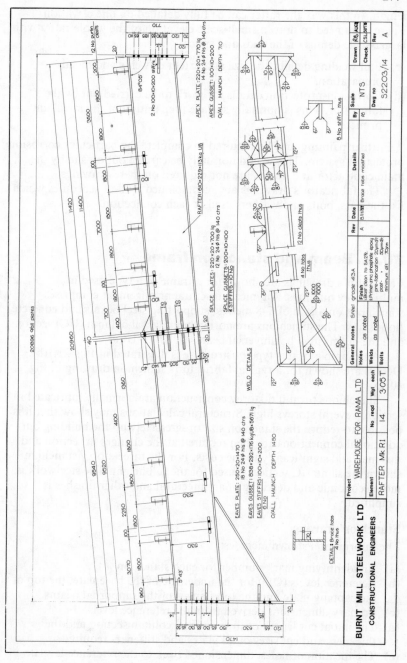

Fig. 30.

length. In this way the rafter can be fully detailed on one elevation. There is no need to draw a small-scale elevation of the whole rafter with larger-scale details of the ends and splice.

(c) Full welding details, the size and extent of welds and metal preparation.
(d) Corrosion-protection details for work to be carried out by the fabricator. This is normally surface preparation of the metal and a primer.

Further painting may be required to complete the particular corrosion protection system. This would normally be carried out on site by the main contractor and details are not required on these drawings.

The fabricator's drawings are accompanied by bolt schedules which list the nuts, bolts and washers used at each connection.

30.3 Beam and stanchion frame

Fabricator's drawings are shown for a frame of a type used in multi-storey buidings. A typical structural arrangement comprises stanchions on a grid of 5–8 metres, supporting steel beams and concrete floor slabs. The stanchions are normally universal column (UC) sections, and the beams may be universal beams (UBs) or UCs.

An example of this type of structure is illustrated in Fig. 21.1(b). Details are shown for a typical fabricator's stanchion drawing in Fig. 30.3.

Stanchions for multi-storey construction are normally fabricated in sections several storeys high. Structural calculations may show that it is possible to reduce the stanchion size progressively up the building but additional connections are then required at the changes in section and this may add significantly to the costs. For the three-storey stanchion shown the same UC section is used for the full height. The steelwork is more economic and detailing of architectural and other finishes is simplified.

Stanchion drawing

The information shown includes:

(a) An identifying mark number for each stanchion.
(b) Reference levels (OD) for the underside of the baseplate, the top of the capping plate and the top flanges of the supported beams.
(c) Primary dimensions derived from the reference levels.
(d) The initial cut length of the universal column section and the thickness and cut sizes of all plates and stiffeners required.
(e) The location of holes, dimensioned back to the reference levels.
(f) Welding details.
(g) Corrosion-protection details for work to be carried out by the

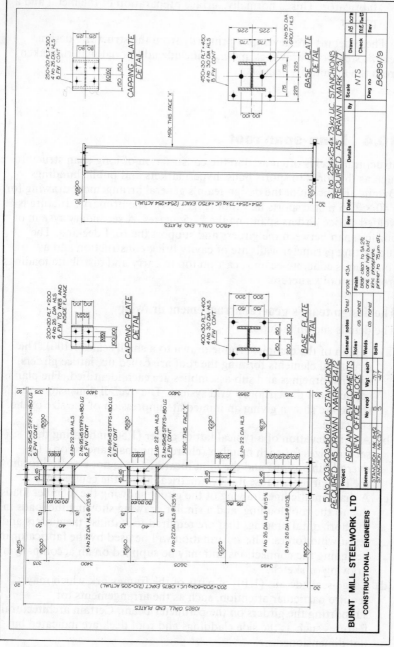

Fig. 30.3

fabricator. This is normally surface preparation of the metal and a primer.

These details are developed primarily from the structural design team's steelwork drawings and may include additional material taken from the architect's details.

30.4   Long-span roof

Structural steel is often used in the construction of long-span structures such as the roofs to sports halls, hypermarkets and public buildings. Details are shown of the design team's general arrangement drawing for a 40 × 27.5 metre sports hall roof (Fig. 30.4). The primary structure is welded lattice girders spanning the 27.5 metres. A secondary system of purlins span between the girders and support the roof decking. The supporting perimeter walls are of cavity brick construction and a continuous concrete beam is cast on top to carry and distribute loadings from the roof structure.

The design team's general arrangement drawing

The details shown include:

(a) A plan of the complete frame drawn to a small scale (1:100). The structural elements forming the roof are called up, lattice girders, purlins, bracings and sub-assemblies are each identified. The plan is set out on a dimensioned grid system. The overall dimensions of the frame are shown giving an immediate impression of the size of the structure.

(b) A part elevation of a typical lattice girder (1:50) indicating the section sizes for each member (chords, internals and so on), the overall dimensions and the joint locations. It is not essential to show the full girder because it is symmetrical about the centre-line.

(c) A diagramatic line drawing of the girder showing the member loads due to the roof loading, and a similar drawing showing the loads in the horizontal bracings that are required to stabilise the girders and resist wind forces. This information will be used by the fabricator in designing the connections and may be supplied on an accompanying drawing or sketch.

(d) Large-scale details (1:10) of parts of the framework which may require particular attention, such as the arrangements for supporting the girders on the perimeter beam. Certain architectural features such as the side claddings and roof deck are indicated but not dimensioned.

(e) 'Notes box' items are included for steel quality, welding and the corrosion-protection system.

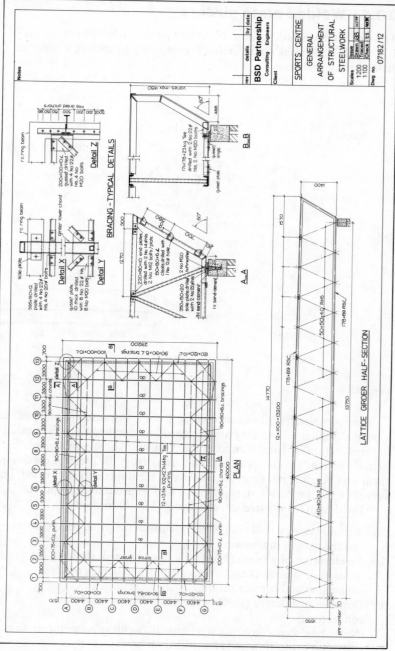

Fig. 30.4

30.5 Welded plate girders

The welded plate girders for a 40-metre-span river bridge are shown in Fig. 30.5(a). The deck construction is completed by casting a concrete slab on the top flanges.

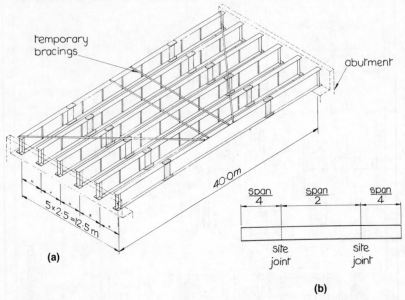

Fig. 30.5

The completed girders are too long to be transported in one piece on a normal vehicle and so they are designed and detailed to be fabricated in three sections (b) and then assembled on site. The site joints are bolted and made at the quarter-points away from the areas of maximum shear and bending moment. Alternatively, the joints may be welded, but very stringent preparation, supervision and inspection are required.

The design team's steelwork drawing (Fig. 30.6) shows:

(a) A small-scale (1:100) plan which identifies the principal elements and the overall dimensions.
(b) An elevation of a beam (1:50) fully dimensioned, showing the location of site joints and web stiffeners and details of shear connectors and camber profile.
(c) Dimensioned details (1:20) of the web stiffeners and shear connectors.
(d) 'Notes box' items covering steel quality, welding procedures, corrosion protection and identification.

A joint load and moment diagram is not included since the design team design the site joints.

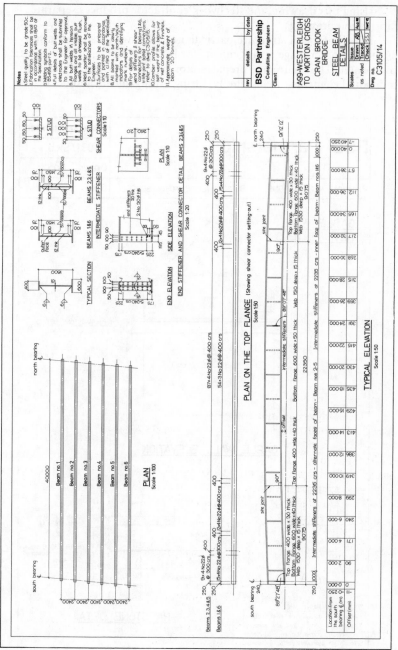

Fig. 30.6

Using this drawing, the steelwork contractor produces separate fabrication drawings.

For this type of structure particular attention is given to the site assembly and cranage arrangements for lifting and placing the beams over water. The contractor prepares fabrication drawings of the individual elements, site assembly, lifting arrangements and details of

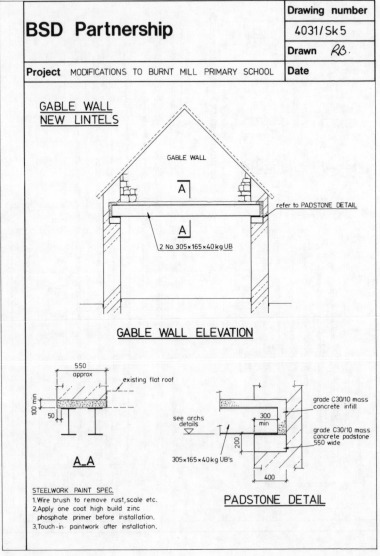

Fig. 30.7

the temporary bracings required to stabilise the beams before the concrete deck has been cast. The contractor may wish to modify the details to accord with particular construction procedures and this would be done in consultation with the design team.

30.6 Steelwork in minor works

The use of structural steelwork in a minor works project is illustrated in Fig. 30.7, which shows twin universal beam lintels required for the structural modifications to an old stone building.

The steelwork is very simple and it is not necessary to prepare a full-sized drawing. Instead, the details are shown as an A4 sketch in the form of an information sheet rather than a conventional structural drawing. The sketch is drawn to scale and critical information is given. This includes the size of the padstone, the minimum length of bearing on the padstone and requirements for the infill concrete between the lintels and the gable masonry. However, stating a precise length for the lintel may not be advisable because work on old buildings is often unpredictable and it is preferable to take site dimensions as the masonry is broken out. These are used to determine the actual lengths required for the steelwork, which is then made to suit.

Work of this type is often carried out by small building contractors who are not familiar with the extensive steelwork specifications required for larger projects. The sketch should be accompanied by a concise specification appropriate to the scale of the work.

Index